THE MANUAL OF

MARINE INVERTEBRATES

MARTYN HAYWOOD · SUE WELLS

THE MANUAL OF

MARINE
INVERTEBRATES

MARTYN HAYWOOD · SUE WELLS

Tetra Press

16038

Above: In this 'living reef' aquarium, intense lighting and a strong water current simulate the bright, fresh conditions these creatures enjoy in the wild. A bright blue damselfish and an orange flame-backed angelfish add lively points of colour against a background of anemones, corals, crinoids and seafans.

A Salamander Book

©1989 Salamander Books Ltd.,
Published by Tetra Press,
3001 Commerce Street,
Blacksburg, VA 24060,

ISBN 1-56465-139-8

All correspondence concerning the content of this volume should be addressed to Tetra Press.

Credits
Editor: Vera Rogers. Designer: Jill Coote,
Color reproductions: Magnum Graphics Ltd.
Filmset: PPC Limited,
Printed in Italy.

THE AUTHORS

Martyn Haywood has pursued a lifelong interest in fishkeeping. Having kept and bred sticklebacks as a child, he went on to keep freshwater tropical fishes and then embarked on setting up marine tanks, which have been a major interest for 17 years. As livestock manager for a major London aquatic business and now as proprietor of his own aquatic centre in Hampshire, Martyn is ideally placed to advise on all the practical aspects of selecting and keeping marine invertebrates. He has written numerous articles on the subject for magazines in the UK and overseas, and in Parts Two and Three of this book combines his skills as author and marine expert to encourage both aspiring and committed aquarists to establish their own invertebrate tanks.

Sue Wells BSc., MSc., graduated from Cambridge University with a degree in Zoology. She has travelled widely, researching, lecturing and writing on invertebrates and coral reef conservation. At the International Union for Conservation of Nature and Natural Resources – the IUCN – in Cambridge, she was co-compiler of the Invertebrate Red Data Book, with special responsibility for 'non-insect' invertebrate groups, particularly molluscs and marine species. She has prepared Part One of this book.

CONSULTANT

Jennifer George BSc., MSc., CBiol., FIBiol., is a Principal Lecturer in Biological Sciences at the Polytechnic of Central London. She has over 20 years experience of lecturing on invertebrates and aquatic ecology to both BSc. and postgraduate students. In adddition to her research on the biological effects and cycling of various pollutants, such as pesticides, metals and hydrocarbons, she undertakes consultancy work and has participated in marine expeditions around the world. Her SCUBA expertise has enabled her to gain first-hand insight into the underwater world of invertebrates and she has acted as consultant on Parts One and Three of this manual.

CONTENTS

PART ONE

WHAT IS AN INVERTEBRATE?

The term 'invertebrate' is a convenient catch-all title, loosely covering all those animals that do not have a backbone (or vertebral column) and an internal skeleton. Invertebrates range in size from microscopic planktonic animals to giant squid, fierce predators that roam the deep, colder waters of the world. There are so many species of invertebrate animals that it is difficult to estimate the total number involved. Of the two million species of animals in the world, about 97 percent are invertebrates. Land-living insects, spiders and worms make up a large proportion of these, but there are probably just as many invertebrate species in the seas as on the land.

The range of marine invertebrates is so vast that there is no sea habitat in which they do not occur. The species found in aquarium shops come primarily from the shallower, warm waters of the tropics, but the cold waters of the poles contain massive populations of shrimps, anemones, sponges and similar creatures.

So adaptable are invertebrates that they have even overturned one of the major, and previously unassailable, scientific precepts, namely that in the first instance all life on earth is primarily dependent upon the sun's energy. (For example, a cabbage captures light energy through photosynthesis and is then eaten by a rabbit that in turn becomes food for a fox.) Now, however, scientists have discovered a small but flourishing ecosystem of bacteria, sponges, molluscs, crabs and filter-feeding worms that exist at depths impenetrable by the sun's rays. The primary energy source for these creatures is submarine volcanic activity in the form of heat and emitted chemicals.

Naturally, these creatures are beyond the scope of the home aquarist, but even so, the invertebrate keeper is faced with selecting and housing creatures with widely different lifestyles and requirements. In Part One, we take a closer look at each phylum (major group) of animals of interest to the aquarist. As well as illustrating and describing the structure and behaviour of species suitable for the aquarium, this overview includes some of the other fascinating animals that make up the diverse world of invertebrates. Clearly, keeping marine invertebrates is not a challenge to be taken up lightly, but the reward is a window onto an absorbing and otherwise often unseen world.

Left: The dramatic 'chimneys' of a tropical tube sponge dominate this varied underwater scene. The sponge expels water and waste material through the exhalent opening at the top of each stack.

PHYLUM PORIFERA

In evolutionary terms, sponges are the most primitive animals likely to be of interest to the marine hobbyist. When alive, they look very different from the familiar dried sponges used in the bath. Unfortunately, many are difficult to transport – exposed to the air they soon die – and only a few are regularly available. However, many species can be found in well-established 'living reef' tanks, where they usually arrive as accidental introductions with living rock, and some of the encrusting tropical species are very attractive. Deep water sponges are generally white, pale yellow or green, but there are a number of brightly coloured species – green, yellow, orange, red and purple – particularly from shallow tropical waters. The purpose of these colours, caused by pigments, is unknown but it has been suggested that they could have a warning function or could protect the sponge from the sun's rays.

At a conservative estimate, there are at least 5,000 sponge species, but 10,000 may be nearer the total number. Most are marine, with just one freshwater family of about 150 species.

Habitat

Sponges are found in all seas wherever there is a suitable substrate for their attachment, such as rocks, shells, corals, plants, boats, pilings, oil rigs and all the other objects that man provides. They are particularly abundant in shallow waters of the continental shelf. In some areas, sponges are so plentiful that they make up 80 percent of all living matter. Commercial bath sponges may occur in large beds, which make them easy to collect. Sponges have been described as living hotels, their chambers providing temporary or

Right: *Verongia* is a common sponge in the Caribbean, with a skeleton of spongin fibres forming a potlike shape. These sponges can attain an enormous size.

Below: *Clathrina clathrus*, a calcareous sponge, begins as a vase shape and then branches to form a complex mass. The branches empty through common openings.

Above: This brilliant red tubular sponge has developed a branching form, which enables it to take in water and expel waste matter more efficiently in quiet waters.

permanent accommodation for a vast number of other organisms, ranging from algae to fish. For example when the 'residents' of some Caribbean loggerhead sponges were counted, there were several thousand shrimps in each sponge. And the sponge threadworm can occur in tens of thousands in a single sponge, making up a significant proportion of its weight.

Although some sponges are eaten by sea slugs, turtles and some fish, many are toxic, particularly if their preferred habitat is in an exposed site. The toxins deter predators, help to keep the sponge free of larvae, etc. and protect the sponge from colonization by corals and other sponges.

Shapes and sizes

Most sponges are irregular in shape; the shapes often depend on the water current, the space available and the substrate to which the sponges are attached. In strong water currents they often grow as rounded or flattened clumps, but in calm water they may take on a branching appearance. A number of species form encrustations over any solid object, taking its shape. Confusingly, sponges of the same species may look very different under different conditions, making identification difficult.

Sponges vary greatly in size, ranging from less than 1cm(0.4in) to approximately 2m(6.6ft) in height and diameter. Deepwater species tend to be particularly enormous, with the largest species being found in the Caribbean and Antarctic. Some may not grow once they are mature, and some Antarctic sponges are known not to have grown for ten years. Given a good food supply in the tank, many small sponges grow quite rapidly.

Structure

Sponges are unlike any other group of marine invertebrates in that they are simply aggregations of cells and have no true tissues or organs. They are also incapable of movement. Not surprisingly, early naturalists assumed they were plants and it was not until the 1800s, when a sponge pumping water was observed through a microscope, that it was finally realized that they are animals.

The name Porifera means 'pore-bearer' This reflects the fact that cells making up a sponge enclose a system of canals and chambers that open to the surface through many small openings, or pores. The simplest sponges are vase-shaped, with a central cavity surrounded by a wall containing the pores. Special cells, known as collar cells, line the inner wall and draw a current of water in through the pores by means of their whiplike hairs, or flagella. The water current supplies the sponge with oxygen and food particles before passing out through the large opening at the top – the osculum – taking waste material with it. Sponges are very efficient filter feeders, the collar cells straining and ingesting bacteria and minute organic particles from the water. Because these particles are too small for many other species to use, sponges are well adapted to living in the nutrient-impoverished waters of the tropical seas.

Larger sponges need a more efficient filtering system to supply all their needs, so the body wall becomes increasingly convoluted. In the more complex species, the central body cavity disappears completely and is replaced with small chambers housing the collar cells. A number of large volcano-shaped processes may develop, each bearing an osculum out of which the current passes. Sometimes, the water current can be detected as much as 1m(3.3ft) above large sponges. The volume of water pumped through a sponge can be remarkable; for example, in one species it has been calculated that a sponge 10cm(4in) high and 1cm(0.4in) in diameter can pump 22.5 litres (about 5 gallons) of water a day.

Sponge cells often lie in a gelatinous medium, supported on a 'skeleton' of spicules or fibrous material called spongin, which is why sponges feel firm to the touch. Sponge spicules, usually invisible to the eye, look like slivers of glass and their different shapes are an important aid in identifying species. If you rub a sponge between finger and thumb, you can feel the spicules, but take care; in some sponges the spicules penetrate the surface and can irritate the skin.

A simple sponge

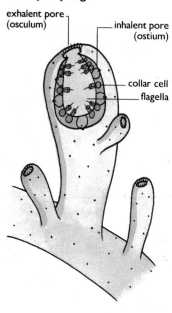

A complex sponge

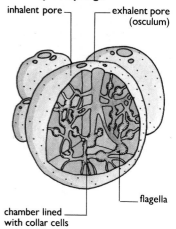

Sponge spicules

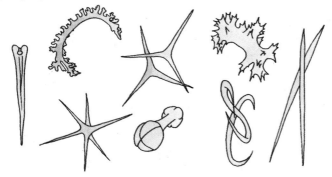

Left: Sponge spicules vary in shape. Calcareous sponges usually have three or four radiating rays; Demospongiae can have knobbed, barbed or star-shaped spicules.

Above: The Venus flower basket *Euplectella aspergillum*, often prized as a decorative ornament, has a latticelike skeleton. The pale body is attached to a rock by a tuft of long spicules and the opening is covered by a 'sieve' of spicules.

Sponge classification

Sponges are divided into four main groups according to their type of skeleton. The Calcarea, the simplest sponges, have calcareous spicules. The Hexactinellida have six-rayed siliceous spicules and include the beautiful deep water glass sponges, such as the Venus flower basket *Euplectella aspergillum*. The Sclerospongiae have massive limey skeletons composed of calcium carbonate, siliceous spicules and organic fibres and can easily be mistaken for corals.

The Demospongiae is the largest group, encompassing species with variously shaped siliceous spicules, a spongin skeleton or a mixture of both. This group includes the sponges of interest to the marine hobbyist (see pages 108-109), as well as the bath sponges and the boring sponges. The bath sponges have been harvested in the Mediterranean for centuries (the Romans used them for padding their helmets, as well as for bathing) and, more recently, in Florida waters. They are allowed to dry in the sun until the soft tissues rot, leaving the spongin skeleton. Boring sponges are in fact quite interesting; their name arises from the fact that they bore through rocks, stony corals and bivalve shells, probably by secreting an acid to dissolve the calcium carbonate. They eventually cause the death of their host and can be quite a pest in commercial oyster beds.

Reproduction

Like the majority of marine invertebrates, sponges reproduce by releasing eggs and sperm which, after fertilization, form larvae that float in the plankton before settling in a suitable place and developing into a new sponge. Sponges have remarkable powers of regeneration; complete new animals will grow from small detached or broken pieces.

PHYLUM COELENTERATA

The coelenterates consist of a vast group of animals, the majority of which are marine. They include jellyfish, sea anemones, sea fans and corals, and are found throughout the world, from the coldest depths to the sun-baked shallows of tropical lagoons. Coelenterates play an immensely important role in marine communities and, like sponges, often provide a habitat as well as food for other invertebrates and fish. Many coelenterates are of interest to the aquarium hobbyist, especially in 'living-reef' aquariums, i.e. those aiming to recreate a section of coral reef. However, it is only fairly recently that many aquarists have succeeded with any but a very limited number of species. In the last ten years, a higher quality of salt mixes, a recognition of the importance of trace elements and stable specific gravities, and improved lighting, filtration and nitrate reduction techniques have all contributed to simplifying the maintenance of corals and anemones in the aquarium. But given that these animals populate such widely different habitats, it is impossible to accommodate them all within one set of environmental conditions.

Structure

Coelenterates seem very variable in appearance, ranging in size from tiny *Hydra* to coral heads (colonies of polyps) several metres across. However, they are all characterized by a radially symmetrical body plan. This means that if you slice horizontally through a coelenterate you will find the body organs arranged in an even circle around a central axis. The body is basically a simple sack, or stomach, with a single opening used both as a mouth and as the exit through which waste is ejected. This is usually surrounded by tentacles armed with tiny stinging cells called nematocysts, used for catching food. Each nematocyst ejects a hollow thread, like a harpoon, into the body of the prey and injects a paralyzing poison. Sessile (non-moving) coelenterates also use nematocysts to protect the living space around them from other animals, including encroaching individuals of the same species.

Polyps and medusae

All coelenterates exist in one of two alternative forms: the polyp, usually attached to rocks or other objects, or the free-swimming medusa. Some groups occur only as polyps and some only as medusae, but some pass through both phases, starting as polyps before budding off medusae as part of their life cycle. A sea anemone is a typical polyp, cylindrical in shape, attached to the substrate at the base end, and with mouth and tentacles at the free end. Polyps often have an external or internal skeleton, as in the corals, and many form colonies by budding off new polyps from the parent polyp. Jellyfish are typical free-swimming medusae, the bell-shaped body having a convex upper surface, below which hang the mouth and tentacles. Medusae are rarely colonial. They swim by alternate contractions of two sets of muscle, which cause them to pulsate. The arms and tentacles often develop into complex shapes to deal with a variety of prey.

Above: *Parazoanthus axinellae*, a zoanthid or colonial sea anemone, has polyps about 1cm(0.4in) high and is often found on rock faces. The tentacles expand to trap food.

Stinging cells (nematocysts)

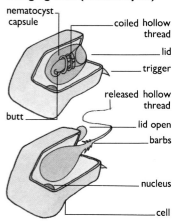

nematocyst capsule

coiled hollow thread

lid

trigger

released hollow thread

butt

lid open

barbs

nucleus

cell

Classification

There are approximately 10,000 species of coelenterate, divided into four main groups: the Hydrozoa, the Cubozoa, the Scyphozoa and the Anthozoa.

The Hydrozoa are mainly marine, but a few species occur in fresh water. The group contains the tiny sea firs (hydroids) and also the complex floating colonies that make up the Portuguese men-of-war. The fire corals that are found on coral reefs and that produce painful rashes if touched, are also included in this group. Hydrozoans characteristically have both polyp and medusa phases.

The Cubozoa are rather similar to jellyfish and include the Australian sea wasps, infamous for their stings that can kill a man within a few minutes. Perhaps not surprisingly, neither of these groups are suitable for the marine aquarium.

The Scyphozoa include all the jellyfish, and the medusa phase is predominant. Jellyfish are well named – even the firmest ones contain 94 percent water. They are highly mobile, and very graceful in the water, swimming by contracting their umbrella-shaped bodies, but they are largely at the mercy of the sea currents. Most species are carnivorous and catch animals with their tentacles, which are armed with nematocysts. They are still virtually unknown in aquariums, perhaps because so many of them are venomous. Tropical species can be extremely dangerous and one jellyfish has tentacles up to 20m (66ft) long. A few species, such as *Cassiopeia andromeda* (see page 125), feed by wafting currents through their mouths and trapping food in strands of mucus. Jellyfish reproduce by releasing fertilized eggs into the sea. These develop into small larvae that float among the plankton in the ocean currents.

A jellyfish

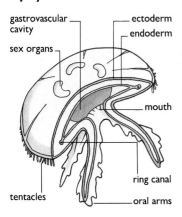

- gastrovascular cavity
- sex organs
- ectoderm
- endoderm
- mouth
- ring canal
- tentacles
- oral arms

Right: The arms of the jellyfish, *Cassiopeia* sp. are edged with frilly extensions. It often settles upside down, enabling the zooxanthellae to photosynthesize efficiently.

The Anthozoa are the largest group of coelenterates. They have no medusa phase and include some 6,000 species. These are of greatest interest to the aquarist, as they include the sea anemones and corals, once known as 'flower animals'. Many corals and anemones are filter feeders or rely on zooxanthellae, but a large number benefit by being fed directly with small pieces of fish or shrimp, sprinkled over the animal or lightly pushed among the tentacles once or twice a week.

The body structure of sea anemones is based on a simple polyp, with multiples of six tentacles around the mouth. In some species, the base is modified for burrowing, but more usually anemones are attached to hard objects by means of a suckerlike disc. Although most seem to remain rooted to the spot, some can move by creeping over rocks. Species of *Stomphia* even leave their rock and swim away when touched by a starfish or predatory sea slug. Tropical anemones are generally larger than their temperate relations and can reach up to 1m(39in) in diameter. Sea anemones are often brightly coloured and a single species may have several different colour forms. Some anemones have zooxanthellae but most catch living prey, including fish, using the nematocysts on their tentacles.

A sea anemone

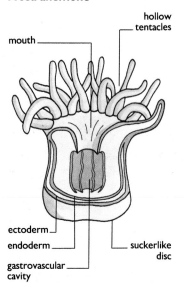

mouth — hollow tentacles

ectoderm — endoderm — gastrovascular cavity — suckerlike disc

WHAT ARE ZOOXANTHELLAE?

Like several other groups of marine invertebrates, many coelenterates have small single-celled plants, or algae, called zooxanthellae that live in the body tissues. Both animal and plant appear to benefit from this 'symbiotic' association, which is particularly common in corals. The coelenterate probably uses the carbohydrates and oxygen produced by the zooxanthellae and the latter use the animal's waste products and assist in the assimilation of vital trace elements from the surrounding water. In the United States and West Germany there is much intensive study into the functions of these algae and the more that is discovered, the more vital they appear to be. Although it has been said that up to 90 percent of their food energy is obtained from these zooxanthellae, hard corals, for example, still need to capture planktonic organisms to survive. The *Tridacna* clams also play host to algae within their tissues but, unlike the corals, these clams will ingest the algae if

they are hungry. With the advent of suitable lighting, it is now possible to satisfy the needs of the zooxanthellae in the domestic aquarium. Generally speaking, coelenterates with beige, brown, green or blue colouring have zooxanthellae and thus require strong lighting of the correct spectrum if these algae are to function correctly; like all

plants they produce food through the process of photosynthesis, for which light is essential. In contrast, coelenterates within the colour range of purple, through red to orange and yellow usually lack zooxanthellae. They are often deep-living or cave animals and generally do not require or, in some cases appreciate, intense lighting.

Right: Bright light is essential if the zooxanthellae in this giant clam's mantle are to flourish.

Above: *Radianthus*, a large reef sea anemone, dwarfs the starfish and fish around it. Many anemones expand at night; others, such as this one, feed during the day.

Others trap organic particles in the water in mucus streams propelled towards the mouth by the tentacles. Several of the large anemones are host to clownfishes, *Amphiprion* sp., that live in their tentacles. The fish are protected by the anemone, and themselves provide protection for their host by deterring predatory fish. They also provide a cleaner service for the anemone. Anemones can reproduce by budding off from the polyp, but they also reproduce sexually by releasing sperms and eggs.

Also of interest to aquarists is a small group of anthozoans that are effectively halfway between anemones and corals. These are the zoanthids, such as the green polyps (see page 125). They resemble small anemones but are colonial. There is no skeleton or basal disc, but the polyps have one or two rings of smooth slender tentacles. Zoanthids encrust rocks and even other animals, such as sponges and corals. The false corals, which include the mushroom polyps (see page 124) are also in a halfway position. Their polyps resemble those of true corals, but have no hard external skeleton. The tentacles often have clubbed tips and are arranged in rings around the mouth.

Anthozoans with eight tentacles

Corals that do not build reefs are classified in another group of the Anthozoa and have an eight-tentacle body plan, unlike the six-tentacle plan of sea anemones and hard corals. This group includes soft corals, sea pens, sea feathers, sea whips, sea fans and also the precious red and pink corals. They are generally colonial, with an internal calcareous or horny skeleton and tentacles that are often branched or featherlike. The precious corals are deep-water or cave-dwelling species and are not suitable for the aquarium, but their beautiful coloured calcareous skeletons are used for jewellery.

Soft corals include the leather corals, pulse corals and cauliflower corals (pages 111-113). They are not as demanding to keep in the aquarium as the hard corals, but still need good water quality and husbandry. The polyps protrude from a fleshy mass, which is sometimes lobed, and is strengthened by calcareous spicules. Deep-water species tend to have more spicules and so are more rigid than shallow-water forms, which are subjected to a greater wave force. The polyps can be completely withdrawn into the body.

Sea pens and sea feathers (see page 110) are fleshy colonies, with short polyps arising from the sides of a central polyp. The lower end of this main polyp forms a stalk that is buried in soft sediments and several species can retract into the mud if disturbed. The skeleton is made of calcareous spicules and may reach 1m(39in) in height. Some sea pens emit waves of glowing phosphorescence when disturbed. They are usually found on the sandy and muddy bottoms of sheltered bays and harbours.

A soft coral polyp

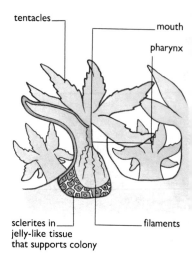

tentacles — mouth

pharynx

sclerites in jelly-like tissue that supports colony — filaments

Below: The rich panorama of a coral reef includes a variety of true corals, soft corals and sea fans. The latter have a fuzzy appearance when the polyps are expanded.

Sea whips and sea fans (pages 114-115) are included in a group known as horny corals, which look rather like plants. The main stem is firmly attached to a hard surface by a plate or tuft of creeping branches. The stem has a central strengthening rod that is generally made out of a horny material called gorgonin (this group is often referred to as 'gorgonians'), although some species have a calcareous skeleton. The short polyps are found all over the branches of the colony but are absent from the main stem. Colonies are often brightly coloured and may reach a height of 3m (10ft). They often provide a home for sponges and hydroids, bryozoans and brittle stars, which stick to their branches.

Hard or stony corals

Many corals have calcareous external skeletons and the biggest group of these – the stony corals – are responsible for building coral reefs. They are most often shades of beige and green, although some are blue or pink. The stony corals are in the same group as sea anemones and have a six-tentacle body plan. The coral animal is basically a tiny sea anemone sitting in a chalky cup, but colonies of these animals can build structures as enormous as the Australian Great Barrier Reef, some 2000km (1260 miles) long and consisting of over 2500 separate reefs. In colonial stony corals, individual polyps may be as small as 5mm (0.2in) in diameter, but in some solitary forms, such as *Heliofungia* (see page 127), the polyp may be as much as 50cm (20in) in diameter. Brain corals (see page 127) are so-named because the polyps are arranged in continuous rows, so that the skeleton has longitudinal fissures in a brainlike mass. The polyps in colonial forms are connected laterally and lie over the

Below: A close-up of a brain coral, *Diploria* sp., shows how the polyps are arranged in rows to form winding valleys; *Diploria* is a Caribbean reef-building coral.

Below: The bright yellow tentacles of the coral *Tubastrea coccinea* only expand at night; by day, it appears a knobby orange or red. It prefers a sheltered habitat.

limestone skeleton that they secrete. Coral reefs are built up over thousands of years; as old die, new colonies form on top.

A coral reef provides a habitat for sea anemones and other corals, as well as for a wide variety of other marine invertebrates, fish and plants. Reef-building corals need warm, clear water – the temperature should rarely drop below 21°C(70°F) – and are easily suffocated by sediment. Coral reefs are often described as oases in an oceanic desert, because the tropical waters in which they occur are very poor in nutrients compared with temperate waters. As a result, if too much food is introduced into a tank, many stony corals will retract their tentacles. Clear water is also needed by the zooxanthellae, on which stony corals heavily depend. To build their skeletons, stony corals need a high pH level and a good reserve of calcium in the water. Without this, they do not flourish and appear to come 'unstuck' from their bases. They can, however, survive in the sea in colder, darker waters, but in such situations their capacity to secrete limestone is greatly reduced and reefs are not formed.

Reef corals grow in many different shapes, depending on their preferred position and water depth. Deeper corals and those in sheltered, still waters tend to form branches, while corals in exposed positions are usually compact. New coral colonies can grow from small broken fragments of larger colonies if the conditions are right; this is one way in which coral reefs recover from damage inflicted by storms and hurricanes. Stony corals also reproduce sexually by releasing sperms and eggs. Recently, it has been discovered that corals on the Great Barrier Reef all spawn on the same night, once a year. The reef becomes covered with a mass of swirling eggs and sperm; the reason for this extraordinary event is not yet clear, but it could be that it confuses predators. The stony corals, including *Goniopora, Leptoria, Tubastrea, Euphyllia, Plerogyra* and *Heliofungia* are described on pages 126–133.

Above: Fish swim over the corals, sponges and crinoids that make up this reef in St. Lucia in the Caribbean. Stony corals provide shelter for many other animals.

Below: *Acropora* shedding eggs and sperm. So many corals spawn on the Great Barrier Reef that the water becomes milky. Larvae drift in the currents before settling.

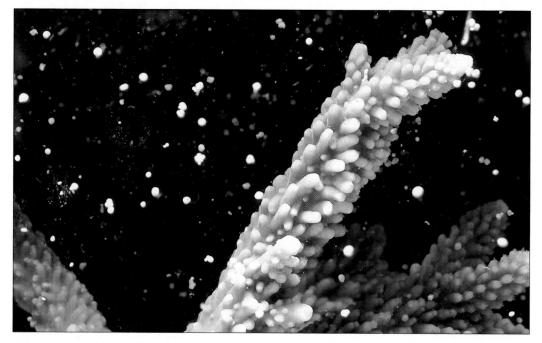

A flatworm

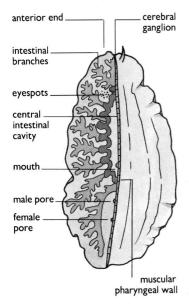

anterior end — cerebral ganglion

intestinal branches

eyespots

central intestinal cavity

mouth

male pore

female pore

muscular pharyngeal wall

Below: *Pseudoceros zebra*, an aptly named and stunning flatworm, found on coral reefs in the Indo-Pacific. It probably feeds at night on other small invertebrates.

PHYLUM PLATYHELMINTHES

The large group of wormlike creatures known as flatworms includes the various parasitic flukes and tapeworms well known to aquarists and studiously avoided by them! However, most fishkeepers would be delighted to come across some of the large and colourful flatworm species found on coral reefs. Unfortunately, these are rarely imported, even though they are fairly common in the wild.

Structure

Flatworms are the most primitive worms and their ancestors occupied a key position on the evolutionary tree leading to the higher animals. The flatworm gut is still a simple, blind-ended tube, and respiration occurs throughout the body surface, but there is an excretory system, muscles along the lines of those found in higher animals and a centralized nervous system, with a tiny brain.

Flatworms are generally small and sombre in colour, and a few are green due to the presence of zooxanthellae. However, the most attractive species – and those of interest to the aquarist – are extremely brightly patterned, often with some form of banding or striping. These species reach about 5cm (2in) in length and are in the family Pseudocerotidae from the Indo-Pacific. This family is in a group characterized by their branching guts. Like most flatworms, the body is flattened, but in this group it is wafer thin and has a leaflike shape, being almost as broad as it is long. They have a recognizable head, usually with numerous pairs of eyes and often a pair of sensory tentacles.

Function and behaviour

The unpleasant flukes and tapeworms belong to the class Trematoda and Cestoda, whereas the free-living flatworms are in the class Turbellaria, of which there are about 4,000 species. Free-living flatworms differ from the parasitic forms in that their body is covered with cilia which, with the muscles, produce the characteristic gliding movement and create the 'turbulence' which has given rise to the group's name. Most of the free-living flatworms are marine and live on the bottom of shallow waters in the intertidal zone, although a few swim freely in the water. They are generally more active at night. The majority are carnivorous. The mouth opens on the underside of the body and sometimes has a muscular tube or funnel-like pharynx that can be extruded through the mouth to grasp or pierce food. The animal has no jaws but digests its prey by releasing enzymes over it and sucking the softened food into the mouth.

Reproduction

Free-living flatworms are hermaphrodite, i.e. they have both male and female organs, but they do not normally fertilize themselves. After copulation between two individuals, a large egg mass is laid from which hatch small larvae. Many flatworms also reproduce asexually by dividing, and freshwater flatworms are noted for their ability to regenerate complete animals from small pieces. It is not known whether this also applies to the tropical marine flatworms.

PHYLUM ANNELIDA

This large group of invertebrates contains the segmented worms, many of which are marine. They have long, soft bodies and are oval in cross-section. The rings on the body, from which the name Annelida derives (in Latin, 'anulus' means a ring), are not merely external, but involve many of the internal organs. These animals have no solid skeleton, but gain rigidity from hydraulic pressure in the fluid-filled body cavity. Alternate contraction of two sets of muscle – one circular and the other longitudinal – allows the animal to move. Most species, apart from the leeches, have bristles or chaetae, protruding from each segment. These also help them to move and may be adapted for other purposes as well. Annelids have a more or less straight gut, with a mouth at one end and an anus at the other. Other characteristics indicate their much higher level of evolution in comparison to the invertebrates discussed so far, namely a good circulatory system and the presence in each segment of excretory organs and a compact mass of nerve cells called a ganglion. There are three main groups of annelids.

Oligochaetes

The class Oligochaeta consists mainly of the familiar terrestrial earthworms, but also includes some important marine species, although none of these are of interest to the aquarist. The diversity and abundance of marine oligochaetes in estuaries and sheltered coasts is only just being realized. Many, such as the tubificids, or sludge worms, are able to tolerate high levels of dissolved organic matter and occur in polluted habitats.

Leeches

Hirudinea – or leeches – are mostly bloodsucking external parasites. There are only a few marine species; most of them are in a single family and feed on fish body fluids.

Polychaetes

The segmented worms of interest to the aquarist are all in the class Polychaeta, the bristleworms. This is the largest and most primitive group of annelids, and the majority are marine. They are often strikingly beautiful and very colourful and, unlike the other two groups of annelids, they show enormous variation in form and lifestyle. Apart from the head and terminal segments, all the segments are identical, each with a pair of flattened, fleshy lobelike paddles called parapodia, which are used for swimming, burrowing and creating a feeding current. The bristles, or chaetae, on the parapodia are immensely variable between species. In the sea mice, for example, they form a protective mat over the back of the worm and give the animal a furry appearance. The bristles of fireworms, on the other hand, are long and poisonous for defence, and are shed readily if a worm is attacked. Fireworms are voracious predators that usually feed on corals, but are known to attack animals, such as anemones, ten times their size. The species *Hermodice carunculata* is sometimes accidentally introduced into the home aquarium; take great care when handling it (see also page 137).

Above: The fireworm, *Erythoe*, from warm tropical waters, is usually found creeping over rocks as it searches for food, protected by its long poisonous bristles.

A fanworm

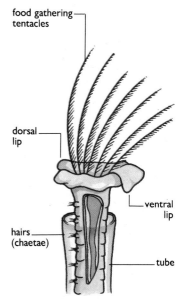

food gathering tentacles

dorsal lip

ventral lip

hairs (chaetae)

tube

22

Polychaetes can be split into two groups on the basis of their behaviour. The errant polychaetes are free-living forms with well developed parapodia used for swimming and, often, for burrowing in the sand and mud. Many of them live under boulders and coral heads, and are common on coral reefs. This group also includes tube dwellers that leave their tubes to hunt for food. By contrast, the sedentary polychaetes are tube dwellers that rarely, if ever, leave their tubes, obtaining food with their tentacles.

In tube dwellers, the tentacles usually form a crown that catches food particles in the water. These are carried to the mouth by cilia on the tentacles. Errant polychaetes often have an eversible pharynx, ending in fierce jaws that seize other animals or suck their body fluids, or grasp large pieces of plant material. The body surface provides sufficient area for respiration in small polychaetes, but larger ones need gills. In tube dwellers, these are usually situated near the tentacles and water is drawn past them by special movements of the body.

Ragworms, which live in the intertidal zone in muddy habitats, are errant polychaetes, and well known to fishermen who often use them as bait. They live in U-shaped burrows, emerging only to feed on plant and animal debris on the surface. Lugworms, also familiar as bait, spend their entire lives in burrows, filtering sediment through the gut and leaving characteristic worm casts on the surface. Both worms respire by directing a current of water through their burrows by regular body contractions.

Below: A sabellid fanworm extending a delicate array of banded feathery tentacles. These are vital for feeding and regrow if bitten off by a fish.

The tubeworms are the only polychaetes of real interest to the aquarist. Both the fan or featherduster worms in the family Sabellidae and the 'Christmas tree' worms in the family Serpulidae make excellent aquarium inhabitants and are recommended for beginners. They are sedentary, spending all their lives in their tubes, and are the most attractive of all the polychaetes.

In tubeworms, the parapodia are degenerate, there are no jaws and the heads are reduced. Instead, feathered, rather stiff tentacles radiate from the head to form an almost complete crown, which is used both as a gill and for feeding. Particles of food are trapped on the branches of each tentacle and channelled to the central rib, from where they flow in a stream of mucus to the mouth. Fanworms have colourful orange, green or purple tentacles forming the crown. Serpulids often have more brightly coloured blue and red tentacles, which may act as a warning device. Fanworms can contract their crowns with startling rapidity, thanks to giant nerve fibres which run from one end of the body to the other within the main nerve cord. The tentacles are extremely sensitive and will respond even to the shadow of a hand passing over the tank.

The cylindrical lower part of the body is protected by a tight tube, secreted by the animal and made of a parchmentlike mucus. Serpulid worms are smaller and produce a tube of a stony calcareous material. As extra protection, a calcareous 'plug' may be present to block the entrance of the tube after the tentacles have been withdrawn. The serpulid group of worms also includes the tiny *Spirorbis* worms, which are often found on rocks and the bottom of boats as little calcareous tubes arranged in a flat spiral about 3.3mm (0.12in) in diameter.

Annelid reproduction
Most annelids have sex cells in each segment; some species are hermaphrodite, but in other species there are individuals of both sexes. At certain times of year, the sex cells are shed into the sea, where fertilization takes place and a ciliated larva is formed. In some cases, the worms die after shedding the sex cells.

Above and below: The jewel-like serpulid Christmas tree worms are commonly scattered over corals on a reef; the main body of the worm remains deep in the coral head, but the two branches of stiff tentacles extend above the surface.

Above: The tentacles, or gills, of *Spirobranchus* tubeworms are used for respiration, as well as for collecting food. The outer tube is permanently attached to one of the stony corals, such as *Porites* or *Diploria*, finger or brain coral.

Reproduction in polychaetes often involves swarming, which helps to ensure that sperm and eggs are released at the same time, rather like the corals on the Great Barrier Reef (see page 20). There is often a special sexual phase that looks quite unlike the normal burrowing individual, with enlarged eyes and parapodia for swimming.

The palolo worm from the South Pacific, though not an aquarium species, is worth mentioning because it is so strange. These worms have sex cells in the rear segments only. These segments change shape and colour and, when ready, this rear section breaks off and rises to the surface, where the eggs and sperm are shed. All the palolo worms do this on one night, near dawn, at full moon in November. When all the segments have risen to the surface, the sea takes on the appearance of vermicelli, turning milky white when the eggs and sperm are released. Meanwhile, the front part of the worm remains in the coral and rocks and regenerates the missing parts. But perhaps strangest of all is the behaviour of the people who live on the islands in this part of the world; the palolo worm is considered a great delicacy and on the appointed day people venture out in boats at dawn and collect buckets of the worm segments as they rise to the surface!

Another interesting polychaete is the fireworm from Bermuda. When the worms come to the surface, the females start to emit a greenish phosphorescent glow. This attracts the males, which dart towards the females, emitting flashing lights at the same time. As the different sexes approach each other, the sex cells are shed.

PHYLUM CRUSTACEA

Crustaceans have aptly been called 'the insects of the sea'.
Although there are some freshwater and terrestrial species, the
majority are marine. They have invaded every possible habitat in
the sea, from deep cold abysses to warm shallows, and have
exploited every way of life in the same way as insects have on land.
There are nearly 40,000 species, ranging in size from microscopic
parasitic and planktonic animals to giant spider crabs from Japan
with a leg span approaching 3m(10ft) and lobsters weighing up to
20kg(44lb). Many species are commercially important, such as the
edible prawns, lobsters and crabs. Others are a major food source
for higher animals. Crustaceans themselves are often active and
efficient predators on invertebrates and fishes.

Structure
Crustaceans have a distinct head, thorax and abdomen, although in
some species the two front sections fuse together. Most have a
telson, or tail piece, and some have a rostrum, or spine, which
projects between the eyes. The number of body segments, limbs
and other appendages varies between species, but there are always
two pairs of antennae on the head, often a pair of stalked,
compound eyes, and at least three pairs of mouthparts. Whereas
smaller crustaceans respire through the body surface, the larger
crustaceans have gills, which are more complex in the more active
species and often associated with the leg appendages. The
appendages of crustaceans show a marked division of labour,
different pairs being adapted for walking, feeding, respiration or
reproduction. The tailfan found in many species – and used for
swimming backwards – is also developed from appendages.

'Suits of armour'
Like insects and the horseshoe crabs, crustaceans have a cuticle of
chitin that serves both as a suit of armour and a skeleton, and it is
this 'crust' that has given rise to the name of the group. In most
large crustaceans, it is strengthened with carbonate and other
calcium salts to keep it rigid. In some species, this carapace, as it is
known, can be thick enough to make the animal almost
invulnerable to all but the most determined predators. The
disadvantage of an exoskeleton is that it must be shed at intervals
as the animal grows. At moulting time, the cuticle splits at the
thorax and the animal squeezes itself backwards, leaving behind a
perfect hollow replica of itself, which is sometimes eaten by its
owner to reduce energy losses. The animal swells rapidly in size
with fluid, so that the new cuticle is larger when it hardens than the
old one. The body is then deflated, leaving space for growth. After
moulting, the animal is soft-skinned and vulnerable; in a tank, be
sure to provide plenty of hiding places for it to retreat into during
this phase. Check that the pH level of the water is not too low,
otherwise the new shell may not harden properly and the animal
will become deformed. Any leg or claw lost before moulting will be
regenerated at the next moult, but make sure that the animal is not
at a disadvantage in the tank and isolate it if necessary (page 97).

Minor crustacean groups

There are eight groups of crustaceans, but not all are considered here. One of the more primitive groups, the Branchiopoda, is largely freshwater, but may be familiar to aquarists because it includes the water flea, *Daphnia*, often used as food for small fish. The brineshrimp, *Artemia*, found in salt pans and ponds, also belongs to this group. The Copepoda include the tiny, usually transparent, herbivores with no carapace that make up most of the sea's plankton and are an important food source for many fish.

Barnacles

It is often not appreciated that barnacles – both the common conical ones and the stalked or gooseneck barnacles, such as *Lepas* (see page 152) – are crustaceans. All barnacles are marine and pass their adult lives attached to rocks or other suitable substrates. These include manmade objects – barnacles are major fouling organisms – and also living animals, such as crabs, turtles and whales. Although they could not look more different from shrimps, crabs and lobsters, dissection shows them to be closely related.

A barnacle has been described as 'an animal standing on its head within a limestone house and kicking food into its mouth with its feet'. It is attached to the substrate by a secretion from its antennae and is enclosed in a shell, developed from the cuticle, consisting of a number of plates. It feeds by circulating water through this and filtering out minute particles with the long, finely branched feet that now act as a filter-feeding mechanism. Goose barnacles are attached by stalks, and are so named because in medieval times it was believed that geese hatched from the egg-shaped shells. This confusion also gave rise to the common name of the barnacle goose! Barnacles have the strange habit of accumulating heavy metals, particularly zinc. Studies in the River Thames in the UK show that up to 15 percent of the dry weight of barnacles may be zinc. Since such high concentrations are easy to measure, barnacles are very useful for environmental monitoring.

Above: *Panulirus versicolor*, the Indo-Pacific purple spiny lobster, has thin spiny claws and graceful antennae. These nocturnal feeders scavenge in caves and crevices.

Left: The shed exoskeleton of a dwarf lobster is a perfect replica of the animal and illustrates how completely the hard chitinous cuticle covers the soft body parts.

Right: The barnacle *Tetraclita squamosa* from tropical reefs has four distinct striated shell plates arranged in a cone. These hide the body and finely branched feet, which are extended for feeding.

Malacostracans

The largest group of Crustacea is the Malacostraca. It contains almost three-quarters of all known crustaceans and many of the most interesting, popular and colourful of the marine hobbyist's invertebrates. It also includes many groups that are not kept in the aquarium, but which are important for other reasons. The isopods are usually flattened and include the familiar terrestrial woodlice, as well as many marine worms, which are usually bottom dwelling and rather dull in colour for camouflage purposes. The amphipods are laterally compressed and curved, with no carapace. The male is often larger than the female and rides on her back before mating. This group includes the sandhoppers that are found almost anywhere in the world on beaches when the tide is out. The tropical mantis shrimps (*Odontodactylus* sp.) have front legs adapted for seizing prey, similar to the terrestrial praying mantises; they lie in wait in front of their burrows for unsuspecting prey – usually fish – before striking. Another important group includes the shrimplike crustaceans of the open sea known as krill, which form the food of the great whales, seals, birds and squid in the southern oceans.

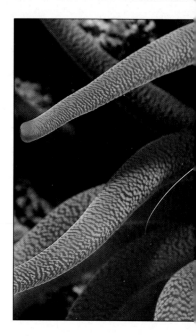

Getting around on ten feet – the decapods

The largest and most familiar crustaceans in the Malacostraca – the crabs, lobsters, crayfish, shrimps and prawns – are all in the Decapoda – literally 'ten feet'. They have five pairs of leglike appendages on the thorax and several rows of gills at the base of the legs, covered by the carapace. Most species are easy to maintain in a domestic aquarium, where they tolerate less than perfect water conditions and accept almost anything remotely edible. They are a particularly interesting group of invertebrates as many are very active. Furthermore, they often display commensal behaviour, which literally means 'sharing food at the same table'. In the crustacean world, it means that they live in close association with other animals for mutual benefit. For example, shrimps and crabs,

A typical decapod crustacean

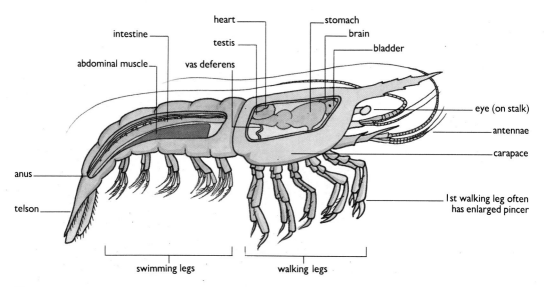

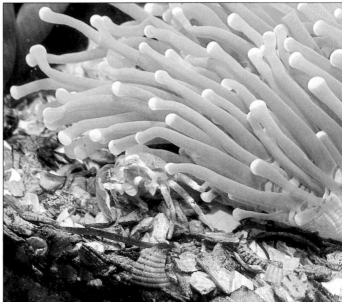

Above: *Periclimenes yucatanicus*, a cleaner shrimp, in the tentacles of *Condylactis gigantea*. Many cleaner shrimps are strikingly patterned; others match their hosts.

Above right: The tiny porcelain crab *Neopetrolisthes ohshimai* lives among anemone tentacles. It has a complicated filter-feeding apparatus at the front of its head.

Below: The cleaner shrimp *Lysmata grabhami*, has a distinctive red stripe on its back. Fish actively seek out the shrimps, which remove debris from their skin.

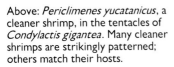

such as *Neopetrolisthes ohshimai* (see page 140) and *Periclimenes brevicarpalis* (see page 150), are often found in the tentacles of sea anemones, between the spines of sea urchins or within the shells of molluscs. They benefit by 'stealing' scraps of their neighbour's food, but may also help in keeping the neighbour and the surroundings clean. The boxing crab, *Lybia tessellata* (page 143), goes one step further and holds anemones in its claws for defence.

Shrimps and prawns

The shrimps and prawns are often laterally compressed, usually with light external skeletons, and include swimming and bottom-dwelling animals. By flexing the abdomen they can move fast enough to escape danger. People often think that shrimps and prawns are different species, but the names have no scientific meaning, although larger species are often called prawns. Some 350 species of this group are used by man for food. They also include the cleaner shrimps, some of which make good aquarium species, such as *Lysmata amboinensis*. Their characteristic coloration may help their 'clients' to locate them. Many other shrimps are suitable for the aquarium (see page 146-152).

Lobsters and crabs

The lobsters and crabs are all bottom-dwelling walking animals. They move on only eight legs, the front pair being modified as huge claws. These are used for seizing and shredding prey, for protecting themselves and to proclaim their territories, as in fiddler crabs (see page 142). The spiny lobsters, such as *Panulirus versicolor* (see page 144), have rather less well-developed but spiny claws. Spiny lobsters often migrate in enormous numbers. Observations have shown, for example, that *Panulirus argus* in the Caribbean may travel as much as 50km (31 miles) in the autumn, covering a distance of 15km (9 miles) each day. Up to 100,000 animals make

the journey travelling in groups, one behind the other, and keeping together by touching and perhaps by using the row of spots clearly visible on the abdomen of the animal in front.

Lobsters with heavy claws are extremely long-lived and some may survive 100 years. They live in holes on rocky bottoms and are mainly scavengers, but will eat live food. Some have an unpleasant tendency to cannibalize their weaker relatives. The larger lobsters have asymmetrical claws; a large one with rounded teeth that is used for crushing, and a smaller one with sharper teeth used for seizing and tearing their prey. The edible parts of the lobster are the well-developed muscles, particularly the ones in the abdomen that are used for swimming. Lobsters are most active at night and their eyes are comparatively poorly developed. Instead, they have sensory bristles all over the body and legs; some of these are sensitive to touch and others are sensitive to chemicals.

Hermit crabs and porcelain crabs
The hermit crabs, porcelain crabs and squat lobsters are all intermediate between lobsters and true crabs and are scavengers. Squat lobsters have large symmetrical abdomens usually flexed below their bodies rather like crabs, and porcelain crabs, such as *Neopetrolisthes ohshimai*, look very like their crab relatives. Hermit crabs, such as *Dardanus megistos* and *Pagurus prideauxi* (see page 142), live in empty mollusc shells in order to protect their soft abdomens. They carefully choose a shell of exactly the right size, and change the shell as they grow. They usually choose right-handed shells, although sometimes they will use the rare left-handed shells. The shell is gripped by their specially adapted legs and its opening blocked by one or more claws.

Right: Fiddler crabs, *Uca* sp., are common in the tropics on muddy and sandy beaches. The large front claw is used for territorial displays outside the burrow.

Left: The colourful spotted lobster, *Enoplometopus occidentalis*, lives in the deeper parts of reefs in the Indo-Pacific and forages for fish and shrimps, mainly at night.

Below: *Dardanus megistos* is a large hermit crab, reaching 20cm(8in) in size. Its strong legs firmly grip its mollusc shell home, which it changes as it grows.

'True' crabs

The 'true' crabs are usually carnivorous walkers on the seabed, although in some species the limbs are adapted for swimming. They have a very reduced abdomen, held permanently flexed below the front segments, which are fused. The carapace is large for the animal's size and extended at the sides. Crabs often have long eyestalks and can move sideways. All these adaptations make the crab a very efficient mover. They are found at all levels, from the deep ocean trenches – where a blind crab preys on the strange animals recently discovered 2.5km (1.6 miles) below sea level around hot vents – to the intertidal zone, beaches and even far inland on large islands. Crabs have a rapid escape reaction and can burrow backwards into mud or sand. Several species of crab, such as *Calappa flammea* (see page 138), cover their carapaces with living organisms, such as algae, sponges and other encrusting animals, to provide camouflage or even a source of food.

Reproduction

Most crustaceans have separate sexes, although some are hermaphrodites. Terrestrial crabs and hermit crabs return to the sea to breed; their mating is usually seasonal, sometimes involving an elaborate courtship ritual. Many crabs mate while the female is still soft from moulting, the fertilized eggs being retained by the female until they hatch, either in a brood pouch or attached to the appendages. The eggs usually hatch into free-swimming larvae that metamorphose through various stages into the adult form. Several stages of larvae look very different from the adult; in fact, one stage – the zoea – looks so different that early naturalists classified it as a completely different animal!

PHYLUM CHELICERATA

King, or horseshoe, crabs, sea spiders and a few mites are the only marine species in this large group that is dominated by the spiders, mites and ticks. The Chelicerata are related to crustaceans and insects, all animals with an external skeleton, or cuticle, made of a material called chitin. The body of a chelicerate has just two sections: the cephalothorax at the front, which includes the head and legs, and the abdomen at the back, which may have some appendages. The name Chelicerata arises from the fact that instead of the antennae found in insects and crustaceans, these animals have a pair of pincerlike mouthparts called chelicerae.

The horseshoe crabs are the only species in this group of interest to aquarists. They are extremely primitive animals; over 300 million years, ago, horseshoe crab ancestors looked much like their modern descendants. They are related to the trilobites, were once widespread and included many species. Today, however, there are only four species, which are considered as 'living fossils'.

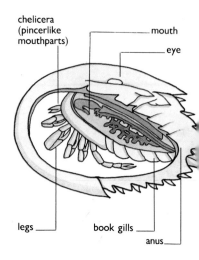

Structure
Horseshoe crabs are light greenish grey to dark brown in colour and can reach a length of 60cm (24in). Males are generally smaller than females. They have a heavily armoured body, and look as if they should be taking part in a science fiction film! The front section (cephalothorax) is covered by a horseshoe-shaped carapace, hinged to the abdomen; the domed shape of the carapace helps the animal to burrow through the mud. Horseshoe crabs have been described as 'walking museums', as the carapace is often covered with large numbers of other organisms, including algae, coelenterates, flatworms, bryozoans and molluscs. The crab has a long mobile tail spine, a telson, which helps it to move and to right itself if it is accidentally overturned. Immature crabs have spines along the top of the telson, but these do not grow and so appear proportionately smaller as the crab gets older; they may help the young crab to burrow, and act as a deterrent to predators.

There are five pairs of walking legs on the front section. The first four pairs have pincer tips and heavy bases that are used for moving food into the mouth. The fifth pair has an even heavier base that is used to crush thin-shelled molluscs, and a whorl of spines to sweep away silt as the animal burrows. The back section of the crab has several pairs of appendages, five of which form gill books – an adaptation unique to horseshoe crabs. These look like the leaves of a book and act as gills for respiration. The gill books are also used for movement by young horseshoe crabs, which swim along upside down.

Behaviour
Horseshoe crabs live on the bottom of sandy and muddy bays and estuaries, and only return to the beach to breed. They generally move along the bottom with a stiff-legged gait, but are capable of swimming in quieter water. They feed on a wide variety of other invertebrates, including worms and molluscs, digging in the mud and passing the food to their mouths with their pincer-tipped front

legs. The American horseshoe crab, *Limulus polyphemus*, can consume at least 100 young soft-shelled surf clams a day and seems able to detect this prey up to 90cm (36in) away.

Reproduction

The breeding period is characterized by the migration of huge numbers of crabs into shallow waters along the shores of bays and estuaries. This is best known in North America, where the massive emergence of horseshoe crabs on the beaches of the Atlantic and Gulf of Mexico is one of the most spectacular phenomena of the coast. The crab makes a useful food for poultry and pigs, a good fertilizer, a bait for other fisheries, and has recently become very important in biomedical research. As a result, large harvests are taken during the breeding period. Thousands of crabs may be taken in each session and, at the peak of the fishery in the 1920s and 1930s, between four and five million crabs were being collected annually, which gives some idea of the enormous numbers that congregate on the shore.

The crabs migrate to the beaches, usually either at full or new moon and within two hours of high tide. The males move sideways to the shore and intercept females heading directly for the beach. The couples then proceed to the beach. The males fertilize the eggs as they are released, and the female lays them in an excavation in the sand, from several hundred to several thousand at a time. The adult crabs leave the beach as the tide ebbs and the eggs hatch in about five weeks. The young crab larvae – closely resembling fossil trilobites – emerge only at an appropriately high tide. Juveniles moult several times a year, burying themselves in the sand at this time to protect themselves from predators; adults, however, may moult once a year or even less.

telson

Below: American horseshoe crabs, *Limulus polyphemus*, mating on the beach. Waders and seabirds take many eggs and hatchlings lying partially hidden in the sand. Adult crabs are protected by the spiny armour plating of their carapace.

PHYLUM MOLLUSCA

The Mollusca is one of the largest phyla in the animal kingdom, with more than 100,000 species. This group has long been important to man, both as a source of food and for its beautiful shells, used for a variety of decorative purposes. Molluscs are found in almost every habitat, and about half are marine. From the aquarist's point of view, many of the most attractive and interesting invertebrates are found within this group. All require good water conditions and, although some should be left strictly to experienced hobbyists, many species are well within the scope of the novice.

Structure

The body of a mollusc consists of a head (although this has virtually been lost in the bivalve group), a muscular foot and a 'visceral mass', which contains the digestive and other organs. The foot is used for gliding over the sea bottom, over rocks or vegetation. In some sea slugs, lobes of the foot are developed for swimming and in a great many species it is used for burrowing. Marine molluscs have gills, which in several groups – particularly the bivalves – are adapted for feeding. Many molluscs have a 'radula', a kind of tongue covered in hundreds of teeth, used for rasping at food. This is made of chitin and is secreted continuously, old rows of teeth at the front dropping off as they become worn.

The name Mollusca means soft-bodied but, like crustaceans, molluscs have an external skeleton to protect and support the body. This is the shell, which is made of a material called conchiolin impregnated with calcium carbonate. Shells come in an infinite variety of colours, patterns, shapes and textures that usually reflect the lifestyle of the animal. They may have regular lines or marks indicating interruptions of growth, which can occur in cold weather. Thus, some species can be aged in much the same way that trees can be aged from their growth rings. The inner layer of shell is often made of tiny blocks of crystalline calcium carbonate and is called nacre or, when the layer is thick, mother-of-pearl. A few species, such as the chambered nautilus and the pearl oysters, consist almost entirely of nacre. Some groups of molluscs have very reduced shells, often internal, or have lost them altogether.

Shells have a variety of functions in addition to acting as a skeleton. Some species, such as the abalones and limpets, have shells that can be clamped tightly to windswept rocks to avoid desiccation when the tide is out, or to avoid being removed by predators. Others have streamlined shells for burrowing. The razor shells are a good example; when alarmed, these can plunge into the sand by as much as a 1m(39in). Perhaps the strangest shells of all are the carrier shells, *Xenophora*. Not content with the natural form of their own shells, they attach stones, bits of coral and other empty shells to the surface, perhaps to camouflage it.

All molluscs have a mantle – a fold of skin enclosing the gills, anus and various other glands and organs. Sometimes it is brightly coloured, perhaps as a warning; at other times it may be coloured to provide a camouflage. The mantle of a cowrie is extended back over the shell when the animal is active and its mottled colour provides

Above: A tiger cowrie with its colourful egg capsules on the Great Barrier Reef. The spotted shell is kept lustrous by the mantle, which expands when the cowrie is active.

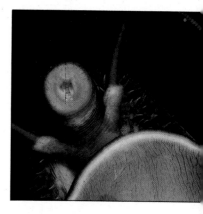

Above: A close-up view of the mouthparts of a tiger cowrie. Cowries are probably omnivorous, using their toothed tongue, or radula, to browse on algae, small invertebrates and fish meat. The tentacles react to light and touch.

good camouflage; it also means that the shell remains shiny and lustrous, unlike many other species whose shells become worn and covered with encrustations over time.

The mantle also secretes the shell and forms what is probably the most famous product of the mollusc – the pearl. Pearls are formed naturally when layers of mother-of-pearl are laid down around particles of grit lodged in the mantle cavity. Some molluscs can be persuaded to form 'cultured' pearls by inserting a tiny hard object, such as a minute piece of shell, into the animal. Commercial pearls come mainly from the pearl oysters, but many people are surprised to discover that other molluscs can form them. The queen conch can produce a delicate pink pearl, although this is not very valuable, and the largest pearl in the world – the pearl of Allah – came from a giant clam and weighs 6.4kg (14lb). The mantle may also secrete poisons for defence, in the form of acids, as in some cowries, or inks as in some sea slugs and the cephalopods.

Reproduction

Marine molluscs either have separate sexes or are hermaphrodites. Apart from the cephalopods and a few other species, such as the Caribbean queen conch, which copulate, most shed their eggs and sperms into the sea at the same time and fertilization takes place in the water. In some cases, the eggs are laid on the bottom or on vegetation in clutches surrounded by a jellylike material. This can be spectacularly colourful, as in some sea slugs. As in many other marine invertebrate larvae, planktonic larvae called veligers develop and float in the ocean currents, often over huge distances. Oysters are known to have been dispersed in this way over 1,300km (780 miles). Some species, such as the giant clams, have shorter planktonic lives and their ranges are therefore smaller. A few species, such as the volutes, have no planktonic stage and produce young that look identical to their parents; these species may have very narrow distributions.

Below: A beautiful spotted sea slug with its ribbon of eggs. The spots may function as camouflage or as a warning, since some species of nudibranch are poisonous.

Classification

Molluscs are an extremely varied group, ranging from small parasitic clams that burrow into the arms of starfishes, to the giant squid that roam the deep waters of the oceans. Of the various groups of mollusc, the three largest are of interest to the aquarist.

Gastropods

The largest group, with 60,000-75,000 species, is the Gastropoda, or univalves. It includes the terrestrial slugs and snails, as well as vast numbers of marine and freshwater species. Gastropods have single shells, often strongly coiled, which usually open on the right-hand side. In species where the adults appear to have differently shaped shells, the juveniles practically always have coiled shells. The shape of the cowrie, for example, is obtained by the final large whorl of the shell enclosing the earlier smaller parts of the spiral within it, as can be clearly seen if an old shell is cut open. Some species have an operculum that operates as a trapdoor to close the shell. This is usually made of a horny material but in some molluscs it is calcareous. The beautiful green operculum of the turban shell is used in jewellery and is known as a cat's eye.

Marine gastropods include herbivores, detritus feeders, carnivores and a few ciliary feeders, in which the radula is reduced or absent. The radula is usually closely adapted to the food that a species eats. The simplest gastropods are the limpets and abalones, both herbivores that use their hard radulas to rasp at seaweeds on rocks. Other herbivores are the cowries, spider shells and conchs. The Caribbean queen conch lives in sea-grass beds and propels itself over the bed using a modified operculum, resembling an outsize fingernail, that it jams into the sand, while the muscular foot heaves the animal forward.

A predatory univalve

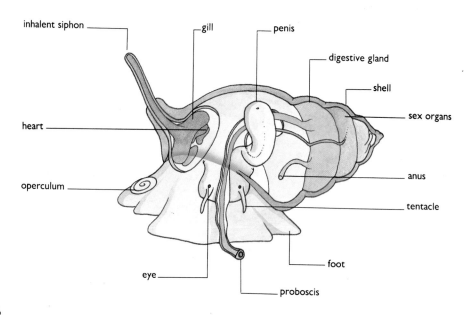

Carnivorous gastropods have fewer, larger, more pointed teeth on a narrower radula than herbivores, and a proboscis that carries the mouth and radula. Whelks feed on dead organic matter, but will also prise open live bivalves by wedging the valves open with the edge of their shells to obtain the flesh inside. Dog whelks bore holes through the shells of other molluscs and barnacles using special teeth and then suck the tissues out. Cone shells have a long proboscis and harpoonlike teeth on the radula. They impale their prey, which includes small fishes, worms and other molluscs, on the radula, paralyze them with a nerve poison and swallow them whole. The poison can sometimes be fatal to man (page 157).

Many marine gastropods are burrowers and have siphons or tubes that extend from the mantle and sometimes the shell. These act as snorkels, enabling the animal to continue to draw in a water current containing oxygen and food into their bodies. The siphons are also used to detect prey from a distance. Many marine gastropods have tentacles on the head, with eyes at the base.

Above: Cone shells are voracious predators. This Pacific species, *Conus purpurescens*, has paralyzed a fish with nerve poison and is in the process of devouring it whole.

Left: *Hexabranchus imperialis*, the colourful Spanish dancer can reach 15cm(6in) in length and swims with a dancing undulating movement, often at night, over coral reefs.

Below: The sea hare, *Aplysia*, a large opisthobranch, releases a purplish poisonous dye from a gland in its mantle when attacked.

Opisthobranchs are one group of gastropods of particular interest to the aquarist. They include the bubble snails that have a very thin, almost translucent shell; the sea hares with a very reduced shell; and the sea slugs, with no shell at all, which can be considered the marine equivalents of land slugs. Some opisthobranchs creep slowly along the sea bottom or over seaweed and corals, but many are agile and beautiful swimmers, such as the Spanish dancer (see page 159) that swims in the surface of the oceans. The opisthobranchs are all hermaphrodites.

Sea hares (see page 158), with their prominent tentacles, are among the largest opisthobranchs and have internal gills and a simple internal shell plate.

Sea slugs are often flamboyantly coloured, either as a warning if they are poisonous, or to camouflage them on the corals and seaweeds on which many species are found. The gills are often in the form of feathery plumes on their backs, and give rise to their other name – nudibranchs, or naked gills. The dorid group of nudibranchs, such as *Chromodoris* (see page 160), have gills in a small cluster at the rear, while the gills of the aeolid group are irregularly placed along the back. Aeolids can withdraw their gills into the body for defensive purposes, while those of dorids are permanently exposed. To counter this weakness, many of the dorids, which feed on anemones and stinging hydroids, 'pirate' the stinging cells (nematocysts) of their coelenterate food and re-use them in their gill tufts as protection against predators. Nudibranchs with smooth or warty backs have no visible gill mechanism and, in some cases, respiration may take place directly through the skin. These species tend to be less attractive than many of the dorids, but often prove hardier and more adaptable to aquarium life. Many also have fleshy extensions of the digestive system on their backs.

A few sea slugs are herbivores, but many are 'grazing carnivores', which may seem a contradiction in terms until one realises that they graze on sedentary animals such as corals, sponges and other invertebrates. Many have distinct dietary preferences and regularly occur in association with certain species. For example, *Chromodoris quadricolor* is found with the sponge *Latrunculia*. Unless the exact food preference is known, keeping these species in an aquarium is very difficult. As a general rule, it is a good idea to house all sea slugs in a well-stocked 'living reef' aquarium in the hope that they will find a suitable food source. Unfortunately, although sea slugs tend to be very attractive, they cannot, with a few exceptions, be recommended to beginners.

Left: The striking black, yellow and white stripes of *Chromodoris quadricolor* are further enhanced by the orange tuft of external gills on this nudibranch's back.

A bivalve

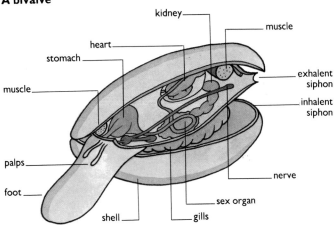

kidney — muscle — heart — stomach — muscle — palps — foot — shell — gills — sex organ — nerve — inhalent siphon — exhalent siphon

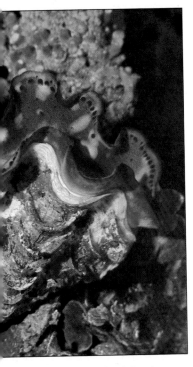

Above: The mantle of the giant clam is so sensitive to light that the valves will close rapidly if a shadow passes over them. In bright sunshine, the mantle is exposed so that the zooxanthellae flourish.

Below: Green mussels are anchored to a rock by their byssus threads. When the valves are open, food is filtered from the water currents that pass over the gills.

Bivalves

The bivalves are the second largest group of molluscs after the gastropods, with 15,000–20,000 species. They include many commercially important species, such as mussels, clams, oysters, scallops and cockles. Their shells, or valves, are in two, usually symmetrical, hinged parts, held together tightly by a pair of powerful muscles. As in the gastropods, bivalve shells can be very variable in shape, colour and texture, the largest being those of the giant clams (see page 163). In bivalves, the head has been lost and there is a pair of large gills, generally used for feeding as well as respiration. These are shaped like leaves or curtains and are covered with cilia that beat continuously to draw in a current. Plankton is trapped by mucus on the gills and carried by the cilia to the mouth. Many bivalves require a high concentration of organic matter in the water and tend to be found in coastal areas. Few can survive long out of water; intertidal species, such as mussels must close their shells tightly at low tide to retain water.

Most bivalves live a sedentary life on or in the seabed. Some, such as mussels and some of the oysters (e.g. the thorny oyster page 165), are attached to rocks and other hard substrates by strong elastic fibres, known as byssus threads. Bivalves feed on filtered phytoplankton or take in detritus with siphons that reach to the surface in burrowing species. Burrowers have a large flattened foot that digs through the sand or mud by a combination of muscle action and blood pressure.

Some bivalves are surprisingly mobile, particularly the cockles, which are capable of leaping, and the scallops, such as the flame scallop (see page 164), which swim by opening and closing their shells and expelling water from the mantle cavity so forcefully that they move under a form of jet propulsion. They have a row of tiny eyes around the mantle edge to detect danger and can therefore make a rapid escape from a predator. The boring molluscs, like the boring sponges, are aptly named for their lifestyle; as soon as the larvae settle, they start excavating into rock or coral, either by using their shell valves, which often have serrated edges, as drills, or by secreting an acid. The shipworms have long cylindrical bodies and bore into timber, using the excavated sawdust as food.

Cephalopods

With 650 species, the Cephalopoda is the third largest group, and includes squid, cuttlefish and octopi. They are the most highly developed molluscs and include some of the most intelligent and fascinating, if also rapacious, invertebrates in the world. The majority are quite large, and many species are totally unsuited to the home aquarium. These include the famous giant squid, which reaches a length of 20m (66ft)! Furthermore, comparatively few species are available with any degree of regularity.

The shell is very reduced or even absent, and a complete shell is found only in the chambered nautilus. However, unlike gastropods, the nautilus only lives in part of its shell, the rest being divided into chambers which are filled with gas and used as a buoyancy organ. Cuttlefish and squid are unusual molluscs in having an internal shell. The flattened cuttlefish 'bone', often found on the beach, is comparatively soft; the squid 'pen' is very thin and reduced.

The mouth of a cephalopod is surrounded by tentacles, derived from the foot found in other molluscs. Octopi have eight tentacles, squid and cuttlefish have ten, and the chambered nautilus has thirtyeight. The tentacles have well-developed senses of touch and taste: those of the octopus (see page 166) are extremely sensitive and can discriminate texture and pattern. All cephalopods are active predators on fish and crustaceans and use their tentacles to locate and capture prey. The mouth has a strong beak for tearing pieces from the prey, which is then pushed into the mouth by the radula. Their voracious appetites are an important consideration when you come to choose suitable tankmates. They are safe with most sessile (non-moving) invertebrates, but will catch and kill any moving animal available. They put a heavy demand on filtration systems, producing a lot of waste products and, at the same time, demanding ideal water conditions. Despite this, they are justifiably popular and can become quite tame.

Below: This cuttlefish has eaten a shrimp, the tentacles of which are still visible. The cuttlefish has ten tentacles; two spoon-shaped ones are visible below the mouth.

A cuttlefish

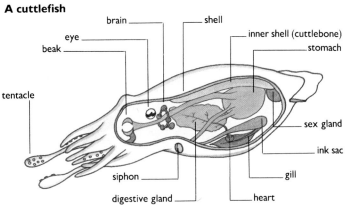

Above: The suckers on the tentacles of this octopus, *Octopus fitchii*, are clearly visible as it crawls over the reef. The suckers are used for locomotion and to locate prey.

Unlike most molluscs, cephalopods are able to move very rapidly. The octopus usually crawls using the suckers on its arms, but it is capable of swimming. In the squid and cuttlefish, the mantle cavity has become a pump that squirts water through a funnel in a form of jet propulsion. The torpedo-shaped squid move around in shoals and reach the greatest swimming speeds of any marine invertebrates. There are even flying squid that can shoot out of the water and glide for some distance, sometimes travelling through the air at 25.5kph(16mph).

The cephalopods have well-developed eyes and the largest brains of all invertebrates. They are responsive to external stimuli – made possible by their giant nerve fibres, similar to those found in some worms – and can respond extreme rapidly to events signalling danger. Octopi even have a memory and can be trained. Many cephalopods are capable of rapid colour change and use this for camouflage, defence and copulation. Pigment cells called chromatophores in the skin expand or contract rapidly, sometimes producing stripes and patterns. By contracting completely they can make the squid almost invisible.

All cephalopods, except nautilus, have an ink sac containing the dark brown ink known as sepia. The cephalopod discharges the sac to produce a 'smoke screen' of ink that hangs like a cloud in the water, fooling the predator while the cephalopod escapes. However smoke screens are of little help to the squid that form the main food of sperm whales; it has been estimated that the whales consume over 100 million tonnes of squid in a year.

Reproduction

Unlike many other molluscs, cephalopods have separate sexes. Octopi go through an elaborate courtship ritual in which the male changes colour and arouses the female by stroking her. Then he transfers a sperm 'packet' on the tip of one of his arms into her mantle cavity, where it fertilizes the eggs. The eggs, as in other cephalopods, are large and yolky and are usually laid in the shelter of a crevice or shell. They are guarded by the female until they hatch as miniature versions of their parents. In some species, the female does not feed during this period of guardianship and dies after the eggs have been hatched. Squid generally lay their eggs in sticky clusters on rocks in open water.

PHYLUM ECHINODERMATA

This group of entirely marine animals consists of about 6,000 species and includes many that are of great interest to hobbyists, such as starfishes, sea urchins, sea lilies, feather stars and sea cucumbers, all of which show a huge variation in structure. However, they share certain constant features. One of the most striking common characteristics is the five-rayed radial symmetry, shown most clearly in the starfish. Like the lower invertebrates, they lack a distinct head, brain and complex sense organs. The nervous system consists mainly of nerve cords along the arms and simple receptor cells over the animal's surface, which respond to touch and chemicals in solution. However, many other aspects of their structure indicate that they are highly evolved invertebrates.

Structure

The skeleton is internal and consists of calcareous ossicles, or plates, that usually bear spines and ridges, from which the name Echinodermata – meaning spiny skinned – is derived. The skeleton is perforated by numerous tiny spaces, which makes it very light while remaining strong.

Echinoderms are also unique in possessing a water vascular system. This consists of five radiating canals containing sea water that connect by side branches to many hundreds of pairs of tube feet. Each tube foot, which in many species has a sucker at the end, can be moved by means of valves and muscles. There is a bladderlike reservoir at the base of the tube foot and when this

Above: The underside of the starfish *Pentaceraster mammillatus* showing the mouth and tube feet. Working in co-ordination, the tube feet creep over the sea bed.

A starfish

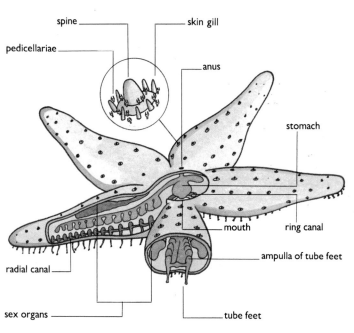

spine — skin gill
pedicellariae —
anus
stomach
mouth — ring canal
radial canal —
ampulla of tube feet
sex organs —
tube feet

contracts, water is forced into the foot and it extends. The tube feet are used in locomotion, respiration, feeding and sensory perception. On its own, a tube foot is a very weak structure, but by working with its neighbours in relays, sufficient pressure can be exerted to enable starfishes to pull apart the two shells of a mussel or cockle.

Starfish and sea urchins have extraordinary tiny protuberances over the body that look like minute tongs or forceps. These were once thought to be parasites on the animals, but are now known to be part of the body and are called pedicellariae. They may be on stalks or attached directly to the skeleton. They have two or more pincers, and their detailed structure is very variable. Their main function seems to be to remove sand and debris from the surface of the animal, but in some species they are used for defence, and may be capable of injecting poison.

Behaviour

The echinoderms are mainly bottom living marine animals. Some are active predators, such as starfishes, which feed on molluscs. Others are filter-feeders, comparable with featherduster worms, that trap small particles of food on feathery appendages. Others survive by sifting through the accumulated detritus on the seabed. A number of species are diurnal, i.e. they are active during the day, but many more confine their activities to the hours of darkness, when their, soft, unprotected bodies are less at risk from predators.

The very bright colouring of the tropical species, and the easy maintenance and often cosmopolitan diet of echinoderms, justifies their popularity among invertebrate keepers. Given good water conditions and a suitable diet, many species can be expected to live in the aquarium for several years. Like many invertebrates, echinoderms often live in close association with other animals. Small starfishes, shrimps and gobies live a well-camouflaged existence among the arms of crinoids, for example, or even inside the bodies of sea cucumbers, other starfishes and sea urchins, emerging only to feed. These 'extra' species are infrequent but welcome bonuses to the aquarium.

Echinoderms reproduce by shedding their eggs and sperm into the sea, where fertilization takes place. Surprisingly enough, the larvae are bilaterally, rather than radially, symmetrical and swim by means of ciliated bands on the body surface. They float in the currents before settling and metamorphosing into an adult.

Feather stars and sea lilies

The free-living feather stars that inhabit shallow seas are most abundant in the tropics. Both they and the stalked sea lilies of deeper waters are in the group known as crinoids. They are the most primitive echinoderms and their history is well documented from the large numbers of fossils, dating back about 500 million years, when crinoids were among the commonest animals in warm shallow seas. Thick beds of limestone in Derbyshire in the UK owe their existence to the accumulation of stalk segments from the ancestors of crinoid species found today. Only the few rare sea lilies now retain this stalk as adults, but the larvae of all modern crinoids are initially anchored to the substrate by a small stem before they break free and drift onto the reefs.

Below: Red feather stars clinging to a rock, having emerged at night to feed. The delicate arms break easily when attacked by predators, but regenerate quickly in the wild.

Crinoid arms are usually forked and branched and their tube feet lack suckers. The tube feet lie in a double row along the upper side of each arm and are used for respiration and feeding, small particles of matter sticking to the mucus on the feet. The sea lilies are fixed to the bottom and look rather like palm trees with upturned fronds. They feed on fine particles sieved from the water.

Feather stars, such as the red crinoid *Himerometra robustipinna*, have a central disc, or cup, from the underside of which grow the short, spiky 'cirri', or hooked appendages, that they use to grip rocks and corals. They can walk rather clumsily on their cirri or swim in a rather spectacular fashion, each arm beating up and down independently with undulations. The cup extends upwards and outwards into five arms that are usually repeatedly branched. These are edged with small extensions called pinnules, and the end result is an animal that looks not unlike a feathery shuttlecock. Their arms have a great degree of vertical flexibility but are capable of very little lateral movement, which makes them very brittle.

Brittle stars and basket stars

Brittle stars, such as *Ophiomastix venosa* (see page 171), and basket stars, such as *Astrophyton muricatum* (see page 172), are in the group known as ophiuroids. Like the feather stars, these have a central disc and their tube feet lack suckers and are used for respiration and feeding only. The skeleton is made up of many ossicles, which fit tightly together. In brittle stars, the disc is flat and the long, thin and very mobile arms are clearly set off from it, unlike those of the starfish. In most brittle stars the arms are smooth, but a number of commonly imported species have short spiky extensions along the arms that may offer some protection from predators. The arms are capable of lateral movements and limited vertical movements, but cannot be coiled around objects. Brittle stars can move remarkably quickly, the arms propelling the animal with snakelike movements.

They feed either by collecting tiny edible particles on their arms as they wave them about, or by tearing off pieces of seaweed or

Above: Colourful starfishes – the spiny red and white *Protoreaster lincki* and the bright blue *Linckia laevigata* – make their way slowly across a coral seabed.

Left: The basket star *Astrophyton* sp. has highly branched arms with tendril-like tips. During the day, it is hidden in a coiled mass, only extending its arms at night to feed.

dead fish. In the sea, they often live in huge beds and can form a seething carpet up to five animals deep. Like feather stars, brittle stars are very fragile, as the name implies, but they are able to regenerate lost arms very rapidly.

In basket stars, the five arms are divided and subdivided to form a greatly branched 'net' and they can be coiled around objects. During the day, they rest on sea fans or rock pinnacles, looking like loose balls of string, but at night, they spread their arms to produce a roughly circular trap, or 'basket', to catch whatever small food items the water currents bring their way.

Starfishes

The starfishes are the most familiar echinoderms and many are brightly coloured. They usually have five or more well-developed, stout arms radiating from the centre of the body. Like brittle and basket stars, these can regenerate relatively easily if damaged. The tube feet are on the underside of the arms and the mouth is in the centre of the underside. Many starfishes feed by everting the stomach through the mouth, encircling and digesting large food items, then retracting the stomach and moving on to pastures new.

Below: The sea urchin *Echinothrix calamaris* moving over *Lobophyllia* coral. Waste material in the faecal sac is being ejected through the anus on the upper surface.

Although not as fast as brittle stars, they can detect food by smell, and crawl towards it at a surprisingly fast rate. Some feed on minute particles but most prey on live animals, forcing open shellfishes or chewing sponges. They can be serious pests of commercial oyster and mussel beds and the infamous crown-of-thorns starfish, *Acanthaster planci*, can wreak havoc on a coral reef if large numbers congregate to feed on the coral polyps.

Sea urchins

The sea urchins and sand dollars in the group known as echinoids have globular or flat bodies with an internal shell, or 'test', of closely fitting plates. At first sight, they appear to bear little relation to the starfishes, but there are clear similarities. They can be thought of as starfish with their arms bent over their backs and their skeletal plates fused together. The tube feet vary in shape and

A sea urchin

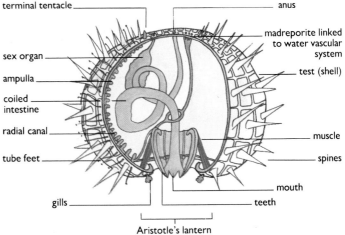

terminal tentacle

anus

madreporite linked to water vascular system

sex organ

test (shell)

ampulla

coiled intestine

radial canal

muscle

tube feet

spines

mouth

gills

teeth

Aristotle's lantern

function, and may be used for movement or creating currents in tunnels. Some urchins burrow in sand and mud, keeping a vertical shaft open to supply water for respiration. Others burrow into rock using the spines attached to the test to dig with. The shape of the spines is adapted to the species habitat; urchins that live on surf-beaten shores have short, stout spines, whereas those from calmer waters, such as *Diadema savignyi*, have longer spines. In addition to providing protection from larger predators, the spines may be used with the tube feet for climbing up rocks.

Many urchins have a unique organ known as Aristotle's lantern, named after its discoverer. This consists of five hard calcareous teeth, suspended from a complex chewing apparatus, and ringing the mouth on the underside of the body. The teeth are used to scrape algal material from rocky surfaces. Sand dollars and heart urchins have no lantern. They burrow in the sand, collecting particles of detritus on their modified spines and tube feet.

Sea cucumbers

The sea cucumbers, or holothurians, are elongated, sausage-shaped echinoderms. The tube feet around the mouth have become sticky tentacles, which vary slightly from one species to another according to the size of the detritus particles on which they feed. Some sea cucumbers sweep the surface of the sand or mud with the tentacles, but others rely on water currents to bring food to them. Other tube feet over the body are used for locomotion. The skeleton is reduced to small crystals embedded in the skin and these produce a rough leathery feel.

When attacked, and possibly as a result of chemical changes in the habitat, sea cucumbers can discharge the stomach and its contents through the anus to help them escape. In the wild, the stomach and intestine are regenerated quickly, but avoid very deflated specimens when buying stock. An alternative defence adopted by the sea cucumber is to squirt out sticky threads. Amazingly enough, sea cucumbers are a gastronomic delicacy in the Far East, where they are known as trepang, or bêche-de-mer, the latter name coming from the Portuguese 'bicho-do-mar' or 'worm-of-the-sea'.

Above: Sea cucumbers use their branched, bushy tentacles to trap planktonic organisms and other foods in the water. The tube feet (here yellow) lie along the body.

A sea cucumber

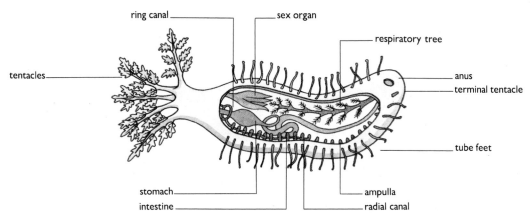

ring canal — sex organ — respiratory tree

tentacles — anus — terminal tentacle

tube feet

stomach — ampulla
intestine — radial canal

PHYLUM CHORDATA

The majority – and the most familiar – chordates are the vertebrates, or animals with backbones. However, there are some species within this phylum that represent the link between the vertebrates and the invertebrates and these are often known as the protochordates. They lack a true backbone, but have a stiff rod, or notochord, in their bodies in at least one stage in their life cycle, and a single hollow dorsal nerve cord. The sea squirts, or ascidians, are the only protochordates of interest to the hobbyist and even these usually arrive in the aquarium by accident.

There are over 1,000 species of sea squirts, many of which live on reefs. Some are solitary and large individuals, up to 50cm (20in) high, while others are colonial and form mats composed of many small individuals. They have a stiff, jellylike or leathery bag-shaped body called a tunic, with large inlet and outlet siphons, although colonial forms have a single communal outlet. Water is drawn in and passes through a strainer, where small particles of food are filtered out before the water is discharged. Their common name comes from their habit of squirting water if they are squeezed.

Sea squirt larvae are like tiny tadpoles and show the typical chordate characteristics. They have a notochord, sense organs and nervous system, all of which are lost when the larva settles and turns into the adult form. Salps are an interesting, free-swimming group of sea squirts that float in the open waters of the oceans. They are jellylike and reproduce by budding, the new individuals often remaining attached to the old ones, thus forming long strings. Many sea squirts are very colourful, such as *Distomus* spp., but their inactivity means that they have never become popular.

The other group of simple chordates contains less than 20 species. These have a notochord and are called lancelets, or amphioxus. They look like little transparent fish, and they burrow in coarse sand in shallow water, lying with their mouths exposed to sieve food from the water current. They sometimes aggregate in large numbers, and are fished and eaten in some parts of the world, particularly by the Chinese.

A sea squirt

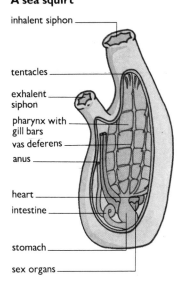

inhalent siphon

tentacles

exhalent siphon

pharynx with gill bars

vas deferens

anus

heart

intestine

stomach

sex organs

Below: This sea squirt has a tough, brightly coloured leathery tunic. The inhalent siphon at one end and the exhalent siphon on the side of the body are edged with yellow.

Right: Diademnid sea squirts grow in these spherical colonies about 2cm (0.8in) high, each with a common, large exhalent opening. Groups of colonies occur on reefs.

PART TWO

KEEPING MARINE INVERTEBRATES

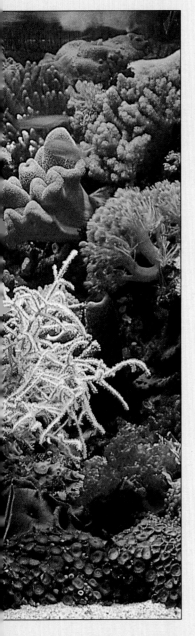

This part of the book explores all the practical aspects of setting up a marine invertebrate aquarium. Obviously, it will be impossible to meet the requirements of every invertebrate within one aquarium, so faced with a new and empty tank, it is a good idea to consider the natural habitat of the creatures you are planning to keep to see how you can best 'recreate' these conditions in the aquarium. *Exploring the options* is therefore the starting point for this part of the book and reflects an ongoing theme: identifying the conditions that prevail in the wild and trying to reflect these in the aquarium. Since the specimens you are housing have been imported directly from the wild, it is clearly an advantage to understand as much as possible about their natural lifestyles.

Once you have some idea of the type of invertebrate tank you wish to keep, the first task is to select an aquarium of the appropriate size, and position it in a suitable spot. Lighting the aquarium correctly is vital and, once again, the natural preferences of the different invertebrate species are the best guide to establishing lighting levels in the tank. In this section we examine the lighting systems available today and how to combine different lights to achieve the desired effect. Installing an adequate filtration system and maintaining optimum water conditions are at the heart of successful invertebrate keeping. These subjects receive full attention in the course of two major sections, with special emphasis on monitoring water quality, a task made much simpler using the reliable test kits available from all good aquarist shops. Carried out regularly, these checks are an invaluable – and frequently underestimated – aid to identifying potential problems before they become real dangers.

Having examined the working principles of aquarium equipment, it is time to install it in the tank. *Setting up* is a step-by-step guide to establishing a fully operational tank, illustrated with some dazzling examples of invertebrate display aquariums. *Selecting and maintaining healthy stock* contains valuable advice on points to look out for, both when buying invertebrates and once they are established in the aquarium. This section includes guidance on treating fish disease in the invertebrate tank. Part Two ends with practical advice on a topic many aquarists find confusing, namely feeding marine invertebrates.

Left: A magnificent 'living reef' tank is the ambition of many aquarists. Advances in modern technology have made it possible to set up such an aquarium, but patience and care are needed to maintain it.

EXPLORING THE OPTIONS

Like all hobbies, keeping fish – and particularly marine fishes and invertebrates – can quickly become a consuming passion. With improved collecting techniques, aquatic shops are able to offer a vast range of invertebrates from widely differing environments. Overleaf are some examples of invertebrates and the range of habitats in which they live. Clearly it is impossible to house them all under the same conditions. At one extreme, for example, are the burrowing horseshoe crabs that live on mud and sand flats, scavenging for worms, algae and small shellfish. In contrast, brightly coloured soft corals require clear, clean water and would not survive in silty water. Some invertebrates are much less demanding than others. Starfish, for example, do not require the high light levels needed by many corals that thrive in the surface layers, while hermit crabs can tolerate much poorer water quality than, say, octopi. Compatibility is another important consideration. Large lobsters and crabs can be extremely destructive, while octopi and their relations will eat any crustaceans or fish that they can capture. Other factors may limit the selection of invertebrates, such as specific food requirements, varying sensitivity to water quality and physical disturbance of the tank decor by one species to the detriment of another.

So start by considering the options. Is this to be principally a fish tank with one or two invertebrates, or would you prefer a specialist tank housing, say, an octopus or a large lobster? Or do you aspire to a complete 'living reef' set-up with a cross-section of living corals, shrimps, worms, anemones and other invertebrates?

Most hobbyists are attracted to marines by the vivid colours of the fishes and then go on to include a few compatible invertebrates. In a fish and invertebrate tank the dilemma that may face the fishkeeper is what to do if the fishes succumb to disease and require medication with copper sulphate. This question is discussed more fully on page 100. The fishes' feeding habits are another important consideration; many will prey on the very animals you wish to keep. A selection of fishes that will safely share an invertebrate tank is included on pages 186-197. The speciality invertebrate tank holds the same appeal for the marine aquarist that large cichlids have for the tropical fishkeeper. Here is a large, often aggressive, but frequently tameable animal that can make a dramatic showpiece. This sort of aquarium is often easier and cheaper to set up than one intended to house a wider range of animals, because the lighting requirements are easier to satisfy. However, in view of the large appetites of the species that inhabit it, a good filtration system is essential to cope with the heavy demands made on it.

The ultimate invertebrate tank has been dubbed the 'living reef' aquarium – an imitation of a shallow-water ecosystem, where the priority is to encourage a wide range of invertebrates to flourish. In this setting, just a few fish add colour and movement. Once established, well stocked and regularly maintained, a living reef aquarium is a spectacular sight, but bear in mind the high cost of setting up such a system. The equipment needed is described in later sections, together with a guide to setting up the tank.

Right: A healthy coral reef teeming with life must be one of the most colourful natural environments. Today it is possible to recreate a small section of it in the aquarium.

Invertebrates: Finding a habitat in the aquarium

Herbivorous sea urchins prefer to scavenge in a shallow water habitat.

A featherduster worm buries its protective tube in gravelly shallows.

The horseshoe crab breeds on shallow sandy beaches.

Soft corals spread their branches in gentle currents to trap planktonic food.

Flatworms glide effortlessly through the shallow waters.

Sea hares browse on the abundant algae that grows in coral lagoons.

Giant clams must have sufficient light to encourage zooxanthellae.

A dramatically coloured starfish scavenges in a sheltered lagoon.

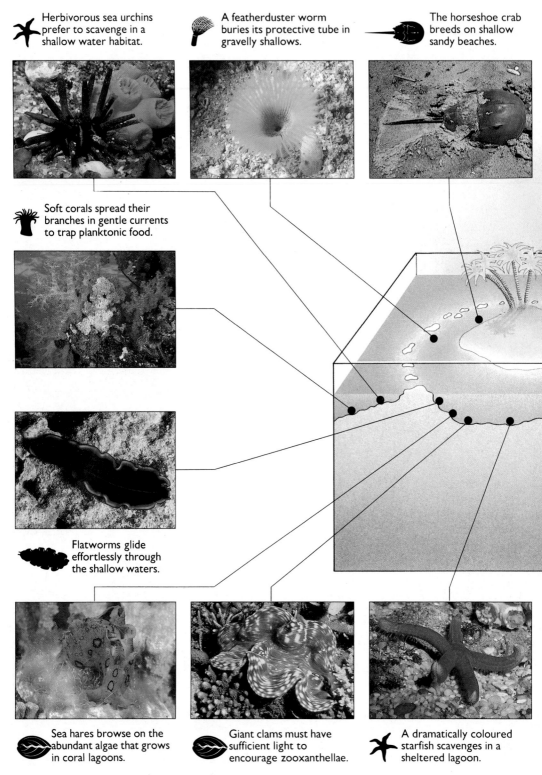

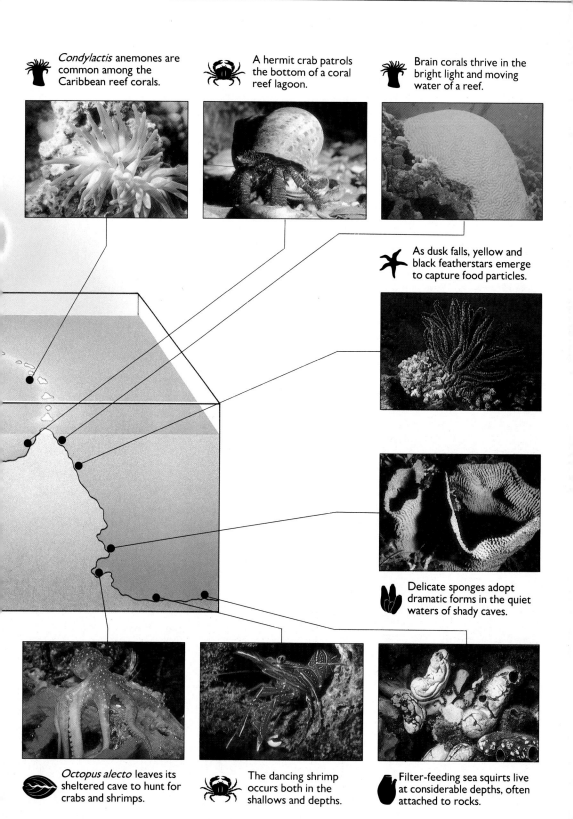

Condylactis anemones are common among the Caribbean reef corals.

A hermit crab patrols the bottom of a coral reef lagoon.

Brain corals thrive in the bright light and moving water of a reef.

As dusk falls, yellow and black featherstars emerge to capture food particles.

Delicate sponges adopt dramatic forms in the quiet waters of shady caves.

Octopus alecto leaves its sheltered cave to hunt for crabs and shrimps.

The dancing shrimp occurs both in the shallows and depths.

Filter-feeding sea squirts live at considerable depths, often attached to rocks.

SELECTING AN AQUARIUM

Choosing an aquarium of the correct size, shape and material is the first consideration when establishing a collection of invertebrates. This early decision can play a crucial role in the long-term success or failure of your venture. The most important thing to remember is that sea water is extremely corrosive to most metals and that aluminium in a marine tank is very toxic. Fortunately, the days of the angle-iron framed tank are long behind us and most aquarists automatically select an all-glass tank made up of sheets of glass glued together with aquarium silicone sealant. The thickness of the glass depends on the size of tank. The advent of these tanks has been the greatest boon to marine hobbyists, as the end result is a suitable container that will not rust or corrode and thus pollute the aquarium water. In North America, aquariums made of moulded or glued acrylic plastic are readily available in a wide range of sizes.

The main drawback of standard all-glass tanks is the fact that most are matched with aluminium alloy lids. If you choose a metal lid for a marine tank, paint it first with a suitable primer and at least two coats of a good-quality gloss paint to reduce corrosion. Use white paint to reflect as much light as possible into the tank. An essential piece of equipment is a tight-fitting cover glass or plastic condensation tray between the tank and lid to reduce evaporation.

Above: Here, a large aquarium has been built into a wall to provide a living picture, well populated with healthy invertebrates. The tank becomes a dramatic focal point.

Left: With the sophisticated tank equipment and good-quality materials available today, it is possible to design a stylish aquarium that will complement the decor of any home.

All-glass tanks are available in an unlimited range of sizes, and many freshwater and marine fishkeepers take pride in stocking large and imposing aquariums. However, unless finances are almost unlimited, there are considerable drawbacks to very large invertebrate aquariums. Apart from the cost of buying the unit, it may be necessary to survey and strengthen floors supporting huge tanks with a capacity of 900 litres(198 Imp. gallons/234 US gallons) or more. (Typical measurements for such a tank would be 200x100x45cm/78x39x18in. In further references to tank capacities, we will quote the equivalent gallonage figures first in Imperial gallons and then in US gallons, separated by an oblique.) When it is in position, filled and stocked, an aquarium of this size will weigh well over 1000kg(2000lbs). Furthermore, large tanks require a correspondingly substantial investment in equipment, particularly in lighting, to produce the intense light that many corals and anemones require.

Small tanks also pose problems. It is possible to maintain invertebrates in tanks measuring 60x38x30cm(24x15x12in) with a capacity of 68 litres(15/17.5 gallons). However, the key to success when keeping any marine animals is to maintain good water quality. Adverse changes can occur very rapidly in a small aquarium, quickly followed by the decline or loss of valuable livestock. Overstocking is always a temptation and a small tank severely limits the aquarist's options when faced with the dazzling array of invertebrates on sale today.

Most beginners to keeping marine invertebrates would be best served by a tank measuring between 90x38x30cm(36x15x12in) and 120x60x45cm(48x24x18in) with a capacity of 150-300 litres (33-66/39-78 gallons). In such tanks you can accommodate a varied range of livestock and maintain fairly stable water conditions. They are reasonably cheap to equip and illuminate with fluorescent tube lighting and have low monthly running costs.

The most attractive tanks are those which form part of a cabinet unit. These not only become a decorative feature in the home, but have the additional advantage of built-in hoods that house several fluorescent tubes and eliminate the risk of corrosion pollution.

Rectangular aquariums of roughly equal width and depth are popular and suitable for marine invertebrates, although cubes can also look very impressive and hold a lot of water in a small space. Think carefully before buying hexagons, octogons and similar odd shapes. They are often much taller than they are long, which makes it difficult to find suitable lighting at reasonable cost. Other standard pieces of equipment often need considerable adaptation before they fit the tank. Furthermore, the owner is faced with vertical seams from every viewpoint, which many people find unpleasant.

Stocking levels

Another disadvantage of very large tanks is the cost of stocking them with sufficient livestock to produce the impressive picture that an invertebrate aquarium can present.

At this point it is worth considering how much livestock you can safely house in a tank of given size. Many elaborate formulae have been put forward over the years to calculate optimum stocking levels. These hinge around tank size, water quality, filtration systems, food inputs and waste outputs and most new fishkeepers find them totally confusing.

With regard to fishes, the simple rule of an absolute maximum of 2.5cm (1in) of fish length per 9 litres (approximately 2 gallons) of water still largely holds good. The situation with invertebrates is somewhat more complex. As living animals they will put some loading on the filtration system, but the range of species available is so huge that it is impossible to say 'so many per litre/gallon'. At its simplest, it is fair to say that animals of different species should not be forced into direct contact with one another. Many corals and anemones expand and contract and you should allow a gap of at least 5cm (2in) between one fully expanded coral and the next. If the animals *appear* to be cramped then you can safely assume that they *are* and the chances of maintaining them are limited.

Siting the tank

Obviously, easy access to a reliable electricity supply is of prime importance. A number of electrical appliances will be in operation and, for safety reasons, you should avoid having cables running long distances to a socket. Ideally, connect the cables to a junction box, or 'cable tidy' and keep them short and neat.

Freshwater fishkeepers have always been told not to site a tank where it will receive direct sunlight, in order to avoid the excessive growth of green algae. However, the situation is not quite the same in the marine invertebrate aquarium. Here, algae growth, be it a decorative 'leaf' species (such as *Caulerpa*), or one of the less attractive unicellular filamentous types, plays an important role in helping to maintain good water quality, since the algae derive much

Above: Anemones should be offered food only when their tentacles are fully expanded. Always remove uneaten food. When a large anemone captures a small fish or shrimp, it quickly enfolds its tentacles around the prey animal, bringing more stinging tentacles into action. Once the animal is killed and fed into the central cavity, the tentacles quite rapidly unfurl to present the beautiful, and seemingly harmless, picture of a 'flower of the sea'.

Above: This aquarium contains much more algae than would appear on a natural coral reef. Nevertheless, the algae is pleasing to look at and plays a beneficial role in water quality management.

of their nourishment from the waste products in the water. Algae are also a very useful food source for many of the animals in which we are interested. Furthermore, many of the corals seem to perform much better if they receive a certain amount of natural light. Generally speaking, choose a position where the tank will receive two or three hours of direct sunlight each day. In an ideal position, a certain amount of algae will form on the front glass of the tank, but this is generally much softer than that in a freshwater aquarium and is easily removed with a magnetic cleaner, or a clean nylon scouring pad. Do not site the tank where it will receive continual sunlight, as algae growth will be too rapid and the glass will probably need cleaning every day.

You must also take into account the weight of the aquarium when it is filled with water. Bear in mind that 1 litre of water weighs 1kg (1 Imp. gallon weighs 10lb/1 US Gallon weighs 8.3lb). Therefore, by the time you have filled a 136 litre(30/36 gallon) tank, equipped it with rockwork and the appropriate substrate and taken into account the weight of the tank itself, the loading on the floor is about 227kg(500lb). However, this figure does not seem so daunting when you realise that it is only about the same as the combined weight of three average people. Most modern houses with concrete floors, or houses with wooden floors in good condition, can support this sort of weight quite comfortably, provided the strain is evenly spread. On wooden floors, ensure that the weight of the aquarium is spread across the maximum number of joists and avoid using metal stands since these have the effect of concentrating four loading points on the floor.

If you use an all-glass tank on a metal stand (but not on a wooden floor) you must first cushion it, unless the manufacturer specifically states that this is not necessary. Place a 2cm(0.75in) thick board on the stand and cover this with a 1.25cm(0.5in) layer of expanded polystyrene on which to stand the tank.

LIGHTING AND HEATING

Choosing the best form of lighting plays a vital part in ensuring long-term success with many invertebrates. Many aquarists come to marines from the freshwater hobby, where the red-enhancing Grolux fluorescent tubes are commonly used. Some will have added further full-spectrum white tubes to give a better visual colour balance to the tank. All, however, will have to rethink their lighting arrangements if they plan to keep a wide range of invertebrates.

Many corals and some molluscs and anemones are largely dependent for food production and waste removal on certain species of algae, known generally as zooxanthellae (see also page 16). These algae live in the fleshy tissues, where they utilize the animals' waste products and provide various foods and oxygen in return. So important are these algae that if they perish then their animal host will, in many cases, also die. The major requirement of these algae is sufficient intensity of light at the correct wavelength.

Penetration of sunlight into sea water

0m

100m

Above: As sunlight penetrates the surface of the water, the various wavelengths are absorbed at different rates. Red light is lost first; blue travels the furthest.

Left: Because blue light penetrates sea water to a greater depth, this pink *Dendronephthya rubeola* appears blue. In the wild, this camouflage provides protection.

Right: Seen under normal aquarium lighting, this closely related alcyonarian coral reveals the intensity of its natural, and very beautiful, red coloration.

Above: In moderately deep water, red animals appear black, because the red light is quickly filtered out. Dwarf lobsters hide more easily from potential predators when their natural colour goes unseen.

Sunlight in the tropics is very intense compared with the more diffused light experienced in the temperate regions. However, in both cases the white light is composed of various wavelengths of light, producing a spectrum of colours from violet to red. Sea water is a very efficient light filter and red light penetrates only a very short distance. Orange and yellow light are the next to be lost, and only green and particularly blue light penetrate much deeper than 4.5-6m(15-20ft). This phenomenon is clear to see when you watch one of the many undersea explorations featured on television. As the cameras venture further down, everything appears to become a deeper blue and the reds and yellows of the animals become deep brown and black. Many invertebrates take advantage of this; in full-spectrum sunlight many shrimps, crabs and corals are intensely red. If this colour were apparent in their natural habitat they would pose easy targets for predators but, because of the depth at which they occur, red appears black and thus these animals are well hidden. Furthermore, many corals, anemones and similar creatures are very sensitive to blue light because successful growth of the vital zooxanthellae is largely dependent on receiving sufficient blue light. A wide range of invertebrates are not particularly dependent on good lighting and they are some of the easiest to maintain. They include the crabs, shrimps and lobsters, featherduster worms, starfishes and sea cucumbers. On the following pages we look at various lighting systems and discuss their relative benefits for those species that do need balanced lighting to flourish in captivity.

Fluorescent lighting

The choice of fluorescent tubes available today is both wide and confusing; many tubes of similar colour are marketed under various trade names. There are several types of full-spectrum white tubes that 'mimic' natural daylight and these are available in a range of sizes to suit most aquariums. White tubes with a so-called 'powertwist' give off somewhat more light than other white tubes of similar wattage. Such 'powertwist' tubes are extremely effective, but considerably more expensive than the alternatives and are available only in a limited range of sizes.

Grolux tubes will be familiar to many fishkeepers, but from the marine point of view their use is almost entirely cosmetic. Most of their output is in the red part of the spectrum and the human eye finds their colour-enhancing effect on animals very appealing. However, red light is of little direct use to marine life. In some circumstances it appears to promote algae growth, but usually encourages the less attractive varieties. Only introduce additional red light to the marine tank once you have achieved the basic white and blue balance.

All these tubes were originally intended for applications other than aquarium lighting. However, a new fluorescent tube has recently been developed specifically as an aquarium light source. This new triphosphor lamp concentrates its light output in the key areas of the spectrum essential for invertebrate well-being and lush marine algae growth. Its spectral distribution takes into account the fact that various areas of the spectrum are absorbed at different rates as light passes through water.

Actinic blue light

The advent of so-called actinic blue fluorescent tubes that produce ultraviolet light has been a major breakthrough; they have proved an economical and easy-to-use source of blue light at the wavelengths necessary to sustain many types of invertebrates. Be sure to select the correct type; actinic tubes are available in two forms; the 03 and 05 ranges. Both give off blue light, but at different wavelengths; for marine tanks you will need the 03 range. These are currently only available in 45cm(18in), 60cm/24in and 150cm(60in) lengths, at 30, 40 and 125 watt ratings. As a general guide, the minimum fluorescent light requirement for light-dependent species is two full-length white tubes and one full-length actinic blue tube. (In real terms, this means that a 60cm/24in tank needs two 60cm white tubes and one 60cm actinic tube.)

Metal-halide lamps

The most intense form of lighting currently available to the fishkeeper is the metal-halide unit. These are extremely powerful and generally prove very successful, particularly if combined with actinic blue lighting. The major drawback is their cost; metal halide lighting is by far the most expensive currently available.

In the United States, many public aquariums are achieving some remarkable results in their 'living reef' tanks, using a combination of metal halide and actinic blue light. The hobbyist can go some way to achieving equally satisfying results with a combination of white and actinic blue fluorescent tubes.

Spectral output of lights used in the aquarium

Balanced daylight tube

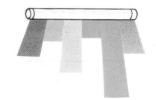

Grolux tube

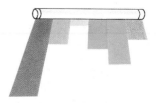

Specialist aquarium tube

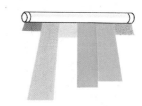

Actinic tube

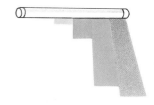

Above: Fluorescent tubes vary considerably in terms of their spectral output at different wavelengths. With careful experimentation, it is possible to achieve a lighting system that satisfies the needs of the fish and algae, as well as the invertebrates, in the aquarium. Metal-halide and mercury vapour lamps produce the very bright light essential for deep invertebrate tanks. They can be very effective when used in conjunction with actinic blue fluorescent tubes.

Right: The high intensity of metal-halide lighting units is clearly illustrated here. The algae and soft corals in this tank require such ample illumination to thrive.

Metal-halide lamp

Mercury vapour lamp

Mercury vapour lamps

A somewhat more accessible alternative is to use mercury vapour lamps. These are considerably cheaper than metal-halide units and, to the human eye, produce a very acceptable light. However, as we have seen, invertebrates tend to be sensitive to different wavelengths. Most lighting manufacturers produce bar graphs showing the light output of their units at specific wavelengths. The bar graph for mercury vapour lights shows significant gaps in the blue-green region, so in order to achieve the best results, supplement these units with additional blue lighting.

If you are proposing to furnish a tank over 45cm(18in) deep and hope to stock a wide range of corals and anemones, you will need to use one or other of these high output lighting systems. Most aquarists opt for shallower tanks, and achieve good results with a combination of suitable fluorescent tubes.

Incandescent lighting

Incandescent lamps and spotlights were once widely used in the aquarium hobby. However, since they are expensive to run, generate a great deal of heat, are shortlived and do not produce light of a suitable colour, they cannot be recommended.

Heating

It is impossible to specify a narrow band of temperature at which all tropical invertebrates will be happy. At one extreme, some animals are found in rock pools and lagoons where the water temperature may regularly exceed 27°C(80°F). On the other hand, some corals, sponges and crustaceans are found in waters more than 9-15m (30-50ft) deep that may be substantially cooler. However, most commercially available invertebrates prefer temperatures in the range 24-26°C(75-79°F). Some corals and anemones are very unhappy at temperatures above 29°C(84°F) and most invertebrates are at their best above 22°C(72°F).

The easiest heating equipment to obtain is the combined heater-thermostat. This submersible immersion heater has a built-in temperature regulator, normally encased in a glass tube, that holds

the tank temperature steady within acceptable limits for each temperature selected. Try to obtain a unit that is calibrated and easy to adjust.

The heating power of a heater unit is rated by its wattage. For example, a 200-watt unit will heat a given amount of water to a given temperature in half the time that a 100-watt unit would need. Generally speaking, most manufacturers recommend too high a wattage for given size aquariums. These days most marine tanks are sited in centrally heated homes or in other locations where background temperatures do not drop dramatically overnight or from one season to another. When a tank is first filled with cold water of perhaps 5-10°C(40-50°F) most fishkeepers expect it to be up to working temperature within 12 hours of switching on the heater. What they do not appreciate is that if a thermostat fails, it will almost invariably do so in the 'on' setting. If this happens, tank temperatures can rise dramatically within a very few hours, resulting in the loss of all livestock. It is much safer to use a lower wattage unit that may take 36 to 48 hours to warm the tank initially. If the thermostat should fail in the 'on' position, more time will elapse before the tank's inhabitants are damaged.

In most situations a 100-watt unit is sufficient to heat up to 136 litres (30/36 gallons); a 200-watt unit will heat 363 litres (80/96 gallons); and a 300-watt unit will heat up to 680 litres (150/180

In an emergency

Cooling tank water
1 Switch off heating circuit.
2 Increase aeration rate to create more turbulence.
3 Immerse external filter return tube in bucket of cold water.
4 Place plastic bags filled with ice cubes in tank.

Heating tank water
1 Carefully reheat some aquarium water in an enamelled pan and return it to tank.
2 Stand bottles of hot water in tank, taking care to avoid an overflow.
3 In the event of serious power failure, lag the tank with expanded polystyrene sheets or wrap it in blankets to conserve heat.
4 Keep spare heater in stock.

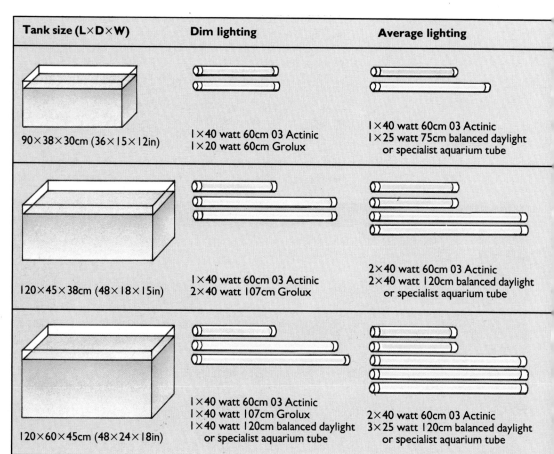

Tank size (L×D×W)	Dim lighting	Average lighting
90×38×30cm (36×15×12in)	1×40 watt 60cm 03 Actinic 1×20 watt 60cm Grolux	1×40 watt 60cm 03 Actinic 1×25 watt 75cm balanced daylight or specialist aquarium tube
120×45×38cm (48×18×15in)	1×40 watt 60cm 03 Actinic 2×40 watt 107cm Grolux	2×40 watt 60cm 03 Actinic 2×40 watt 120cm balanced daylight or specialist aquarium tube
120×60×45cm (48×24×18in)	1×40 watt 60cm 03 Actinic 1×40 watt 107cm Grolux 1×40 watt 120cm balanced daylight or specialist aquarium tube	2×40 watt 60cm 03 Actinic 3×25 watt 120cm balanced daylight or specialist aquarium tube

gallons). Only if the tank is sited in a particularly cool position will you require more powerful heaters. In view of the value of marine livestock, it is well worth using matched, lower wattage units to achieve the required temperature. Using two units provides a kind of 'fail-safe' system; if one unit fails then the other acts as a backup.

A thermometer is essential to enable you to check the temperature of the aquarium, but do not opt for one of the very accurate mercury units. They are fragile and if they break in the tank the mercury will quickly poison the water. Unfortunately, neither the liquid crystal, digital stick-on thermometers, nor the cheap alcohol thermometers are particularly accurate, but they do enable you to see straight away whether the temperature is within the correct range. A steady temperature is more important than one fixed precisely at, say, 26°C(79°F). For more accurate information, use a good thermometer, such as the alcohol-filled units used by amateur photographers for home printing and developing.

Check the temperature of your aquarium twice a day before you switch on the lights and immediately before you switch them off. Some of the very powerful light systems now available to marine hobbyists can raise the tank temperature by several degrees during the day. Room temperatures may also have an influence. This is not usually significant, but occasionally you may need to devise a fan-driven cooling/ventilation system.

Below: Thermometers clockwise from top left: two LCD external; one free-floating or captive design; a simple spirit captive type.

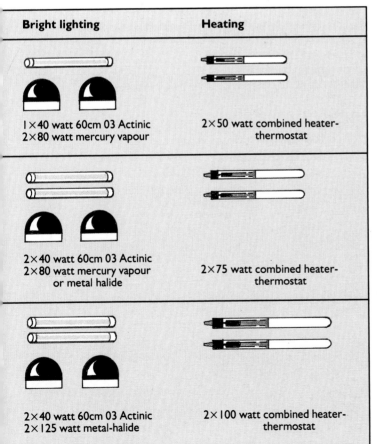

Bright lighting	Heating
1×40 watt 60cm 03 Actinic 2×80 watt mercury vapour	2×50 watt combined heater-thermostat
2×40 watt 60cm 03 Actinic 2×80 watt mercury vapour or metal halide	2×75 watt combined heater-thermostat
2×40 watt 60cm 03 Actinic 2×125 watt metal-halide	2×100 watt combined heater-thermostat

FILTRATION

The most critical and, to the beginner, most confusing aspect of the marine aquarium is how to provide adequate filtration. The function of filtration is to keep the water clean, both visibly and chemically, so that the animals can continue to live, hopefully grow, and possibly reproduce within the aquarium. Filtration breaks down into three convenient, though not mutually exclusive, headings: biological, chemical and mechanical.

Biological filtration

All animals produce nitrogenous waste matter that is poisonous to them, and in the aquarium these wastes appear primarily in the form of ammonia (NH_3). Biological filter systems convert these dissolved waste products into progressively less toxic substances by biological (bacterial) action. In fact, a good biological filter system is at the heart of every successful marine aquarium. Such filters can range from simple in-tank devices to complex units incorporated into 'total' tank management systems. Here, we consider two types of biological filters widely used in marine tanks.

Undergravel filters

In essence, an undergravel filter is a perforated plate, covered by a 7-10cm(2.75-4in) layer of suitable substrate ideally, equal quantities of coral gravel, coral sand and a commercial buffering material. In a traditional 'down-flow' setup, polluted, but oxygenated, water is drawn downwards through the substrate, where colonies of nitrifying bacteria living on the surface of the substrate particles break down nitrogenous wastes in two steps.

Right: In the natural world, harmful ammonia and nitrite are continually converted into less dangerous nitrates, which are in turn used as food by plants and algae. Many invertebrates are extremely sensitive to high levels of nitrite and even nitrate, and efficient biological filtration is essential for their continued survival. Powerful and efficient water pumps will create well-oxygenated water, thus encouraging nitrifying bacteria to thrive in the aquarium.

Below: In a newly established aquarium, the levels of ammonia and nitrite form overlapping peaks as the bacteria in the biological filter multiply and begin to break down these toxic substances.

The maturing process of a biological filter

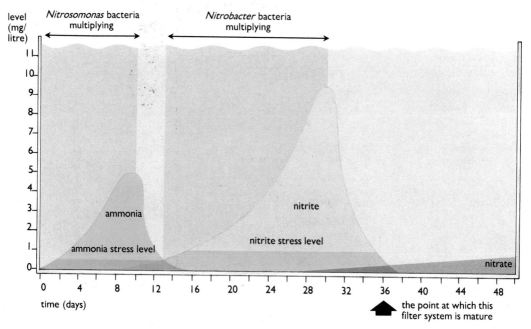

level (mg/litre)

Nitrosomonas bacteria multiplying

Nitrobacter bacteria multiplying

ammonia

ammonia stress level

nitrite

nitrite stress level

nitrate

time (days)

the point at which this filter system is mature

The nitrogen cycle

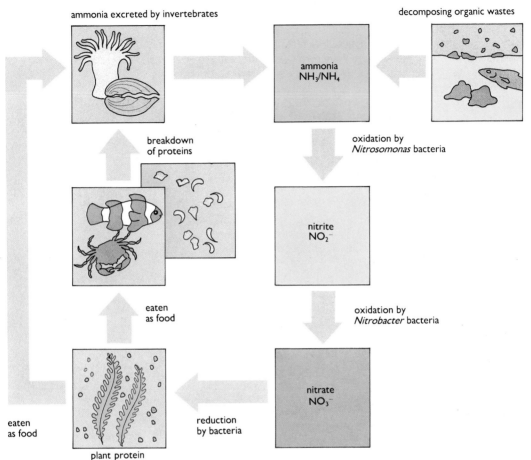

ammonia excreted by invertebrates

decomposing organic wastes

ammonia
NH₃/NH₄

breakdown
of proteins

oxidation by
Nitrosomonas bacteria

nitrite
NO₂⁻

eaten
as food

oxidation by
Nitrobacter bacteria

nitrate
NO₃⁻

eaten
as food

reduction
by bacteria

plant protein

Nitrosomonas bacteria oxidize the highly toxic ammonia (NH_3) to almost equally poisonous nitrite ions (NO_2^-), and Nitrobacter bacteria oxidize nitrites to the much less harmful nitrate ions (NO_3^-). (This nitrification process occurs in nature as a crucial part of the so-called 'nitrogen cycle').

It is vital that this conversion process occurs efficiently because ammonia and nitrites are extremely dangerous, even at very low levels. The nitrifying bacteria are aerobic (oxygen-loving) and so it is important to keep the water moving and well-oxygenated. Water movement can be achieved in two basic ways: by using air pumps or water pumps. Most newcomers to the marine hobby will have arrived via tropical freshwater fishkeeping and will be familiar with air pumps. These can be used to drive undergravel filters, but they are by no means completely satisfactory. The main drawback is that the flow from an air-operated undergravel filter is not strong enough to produce the powerful water currents required by many of the most commonly kept invertebrates. Furthermore, dry salts tend to accumulate at the tip of the air tube connection to the undergravel filter. This reduces the flow of air and in turn reduces the rate at which water is pulled through the filter. In extreme cases it may cause the complete failure of the biological filter system.

A much better alternative are submersible water pumps, known as powerheads, that draw water through the substrate. These small pumps simply sit on top of the undergravel filter uplifts (see page 84 for more information on installing filters). Even the smallest units move a great deal of water and generate beneficial currents. Most models also have a venturi facility, which allows additional air to be drawn into the aquarium to improve the aeration still further. Not least of the benefits of powerheads is their comparative silence in operation, compared to air pump operated filters (provided the air inlet is valved to regulate the influx of air to venturis).

In down-flow undergravel filters there is a tendency for debris to build up within the substrate and effectively clog the system. Using the alternative 'reverse-flow' approach helps to keep the substrate clean by first straining the water through a power filter and then pumping it under the filter plate and thus upwards through the substrate into the body of the tank. This has the added advantage of keeping food particles in suspension, which will benefit many invertebrates. When carrying out a partial water change, use a gravel cleaner to remove debris and agitate the gravel, thus preventing the development of channels in the filter bed.

The nitrates produced as end products are much less poisonous than ammonia and nitrites, but at levels of more than 20mg/

Downflow biological filtration system

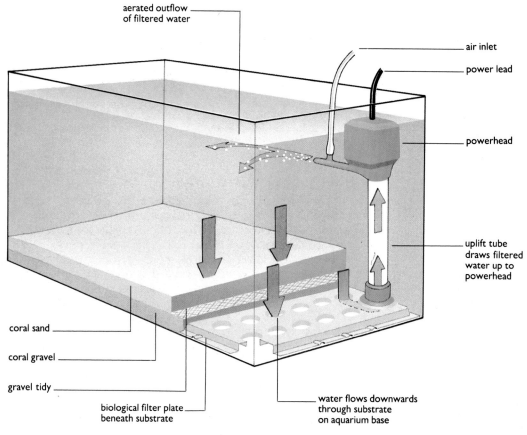

aerated outflow of filtered water

air inlet

power lead

powerhead

uplift tube draws filtered water up to powerhead

coral sand

coral gravel

gravel tidy

biological filter plate beneath substrate

water flows downwards through substrate on aquarium base

litre(ppm) many invertebrates will suffer. You can reduce the nitrate level by carrying out partial water changes at regular intervals and, to a certain extent, by stimulating heavy growths of algae, which use nitrates as a fertilizer. The algae is then harvested from the tank by browsing invertebrates and fishes. (Keep an eye on algal populations, however; they can suddenly die back, causing oxygen depletion in the water. Some algae can also produce toxic substances.) A unit containing nitrate-removing resins to reduce the nitrates in tap water is available from good aquarist shops. Reverse osmosis units, producing pure water, will tackle local pollutants in tap water, e.g. herbicides and pesticides.

Trickle filters

A more efficient system of biological filtration is achieved using so-called 'trickle filters'. The development of the nitrifying bacteria is limited by three main factors: their supply of food; the surface area for them to colonize and, most critically, the amount of oxygen available to them. Undergravel filters meet the first two demands well, but the oxygen-carrying capacity of water is limited.

Dry trickle filters overcome this problem by removing the filter bed from the tank into a separate container, either above or below the tank. Water is pumped or taken from the tank by gravity, and

Reverse-flow biological filtration system

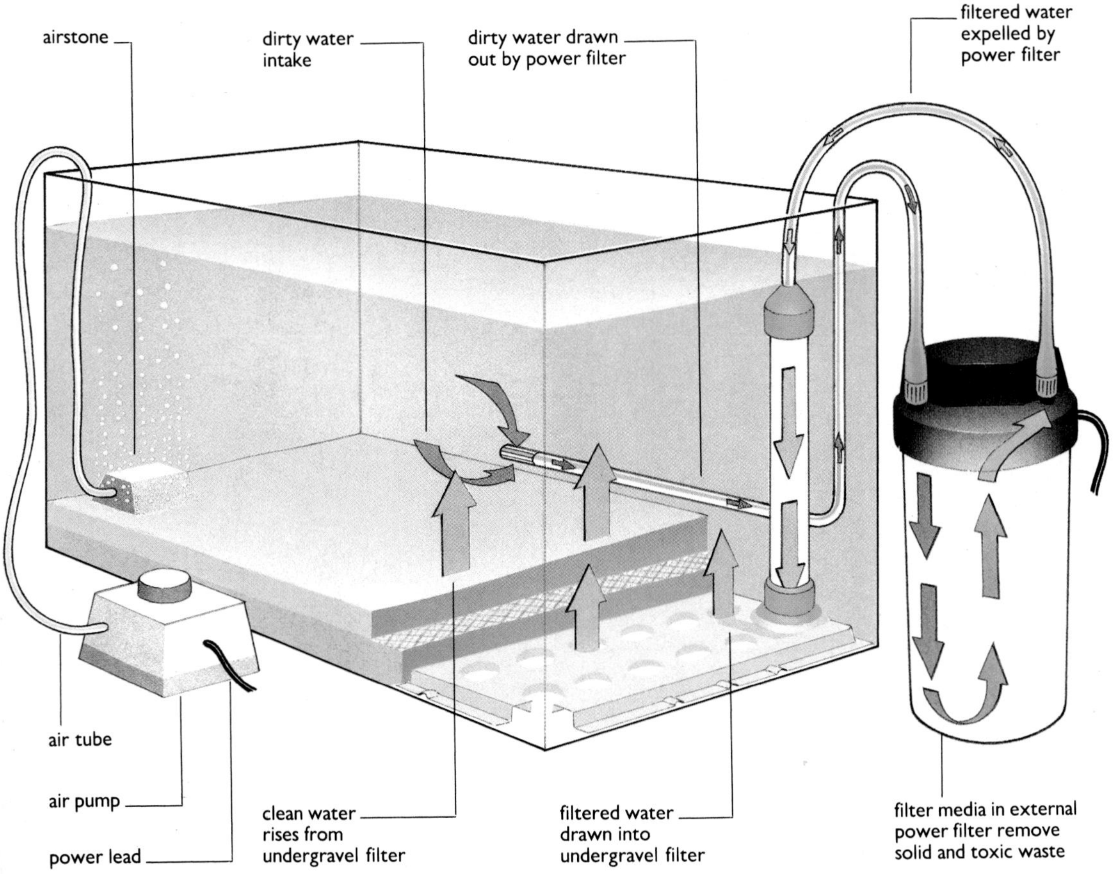

airstone

dirty water intake

dirty water drawn out by power filter

filtered water expelled by power filter

air tube

air pump

power lead

clean water rises from undergravel filter

filtered water drawn into undergravel filter

filter media in external power filter remove solid and toxic waste sprayed over the substrate. The beauty of this system is that the nitrifying bacteria have a virtually unlimited supply of oxygen and thus develop at a prodigious rate. Correctly managed, a dry trickle filter substrate (porous clay granules or similarly inert material) has a biological efficiency at least 20 times that of a similar volume of substrate in a submerged undergravel filter.

Trickle filter systems are available commercially, but keen hobbyists will find it a simple task to set up their own. Remember to pass the water through an outside power filter first, before it flows into the trickle filter. Such mechanical straining will help to keep the filter medium clean and effective for a much longer time.

So-called 'wet' trickle filters provide a way of removing nitrates from the aquarium. At their simplest, these consist of a very porous filter medium kept submerged in a box or canister into which a very slow trickle of water enters from the aquarium. The water is ultimately pumped back into the tank. Because of the very slow flow of water through this unit, oxygen is in very short supply. Anaerobic denitrifying bacteria develop here and obtain their necessary oxygen by breaking down nitrates (NO_3^-), first to nitrous oxide (N_2O) and ultimately to free nitrogen gas. Very few wet trickle filters are available commercially, other than those included as part of complete tank and filter system packages. Very simple units that consist of a box filled with a suitable medium and buried beneath the tank gravel perform a very useful function along similar lines and at very low cost.

Trickle filter

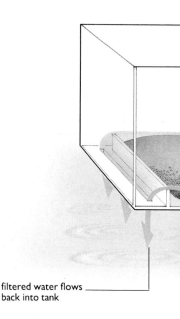

filtered water flows back into tank

Nitrate reduction system

nitrogen gas bubbles released into tank

filter box buried beneath tank gravel

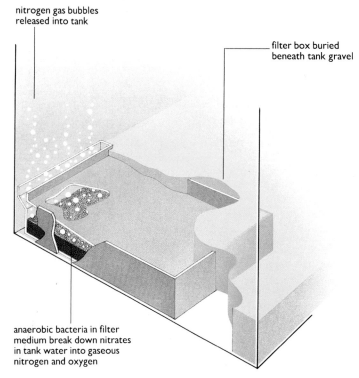

anaerobic bacteria in filter medium break down nitrates in tank water into gaseous nitrogen and oxygen

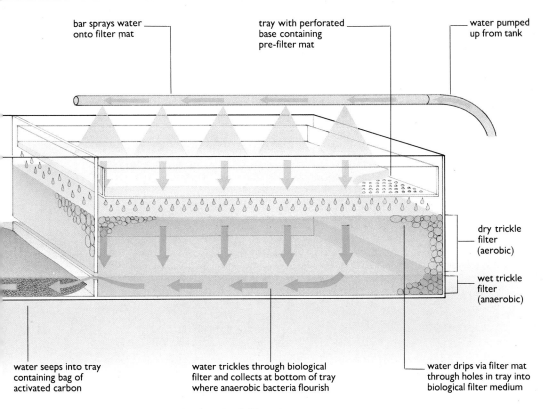

bar sprays water
onto filter mat

tray with perforated
base containing
pre-filter mat

water pumped
up from tank

dry trickle
filter
(aerobic)

wet trickle
filter
(anaerobic)

water seeps into tray
containing bag of
activated carbon

water trickles through biological
filter and collects at bottom of tray
where anaerobic bacteria flourish

water drips via filter mat
through holes in tray into
biological filter medium

Below: Sea whips, such as this
orange gorgonian are sensitive to
nitrate pollution. Monitor the
aquarium water to prevent a
build-up of harmful substances.

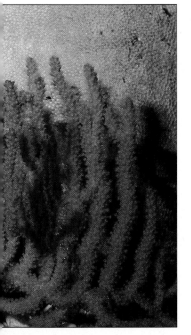

Chemical filtration

In addition to ammonia and related nitrogenous compounds, other
chemicals may need removing from the water – whether they are
produced naturally by the occupants, added as remedies or infiltrate
the tank as pollutants. Since some of these chemicals are not easily
removed by undergravel filtration, it is here that various forms of
chemical filtration come into their own. Of most use to invertebrate
keepers are the specially treated synthetic pads and marine-grade
activated carbon. Both remove chemicals from the water by a
process of 'adsorption', by which the molecules are held by loose
chemical bonds to the surface of the medium. To be effective,
therefore, these media need to present a very large surface area for
adsorption. Both synthetic pads and activated carbon are normally
used in the canister of an external power filter. Carbon is somewhat
cheaper, but an adsorbing pad has the added advantage that its
useful life can be judged visually – the pad changes from white to a
blackish brown when it has adsorbed all it can. In contrast, the
effective life of carbon is not easy to estimate and, once 'full' with
adsorbed chemicals is liable to release them back into the tank, thus
negating all its previous beneficial work. However, used in small
amounts and changed weekly, carbon can be extremely beneficial.

Sea water is a very complex chemical solution containing many
trace elements and organic substances, both beneficial and
detrimental to its inhabitants. It is very difficult to manufacture a
chemical filter that removes only the detrimental molecules while
leaving the others available to the marine animals and plants. This
has led to much, sometimes heated, debate regarding the value of

chemical filtration in the marine aquarium. Generally speaking, if supplementary trace elements, buffer solutions and vitamin additives are used regularly (i.e. to replace those filtered off), then the benefits of chemical filtration outweigh its limitations.

Mechanical filtration

Mechanical filtration serves the simple function of removing visible particulate matter from the water and delivering it to a point where it is easily removed and/or where it can be broken down into less toxic end products. Most filtration media exert a mechanical straining function, but those used specifically for mechanical filtration include filter wool, nylon scouring pads and ceramic pieces. These media are often packed with other materials in the canisters of internal or external power filters.

External power filter

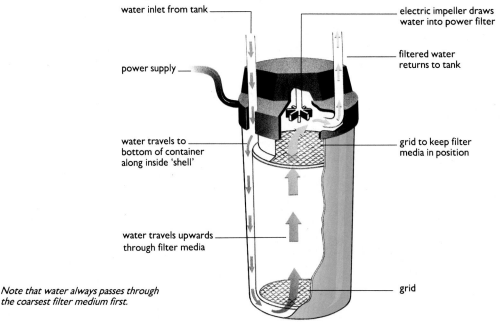

water inlet from tank

electric impeller draws water into power filter

power supply

filtered water returns to tank

water travels to bottom of container along inside 'shell'

grid to keep filter media in position

water travels upwards through filter media

Note that water always passes through the coarsest filter medium first.

grid

Mechanical filter media

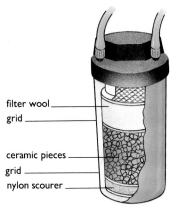

filter wool
grid

ceramic pieces
grid
nylon scourer

Mechanical/biological filter media

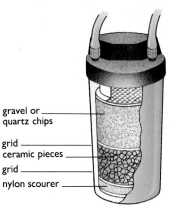

gravel or quartz chips

grid
ceramic pieces
grid
nylon scourer

Chemical filter media

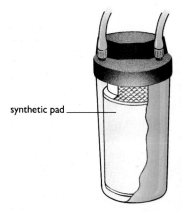

synthetic pad

Protein skimmers

Air-stripping devices, known as protein skimmers, are now widely used and appreciated by marine hobbyists the world over. Many harmful waste products, particularly proteins, phenols and albumens, are easily and efficently removed by these skimmers. In the simplest air-operated units, a column of enclosed water is fiercely aerated with very fine bubbles. The proteins and other wastes 'stick' to the bubbles (a natural tendency of so-called 'surface-active' dissolved organic molecules) and are carried to the top of the column in a stiff foam that is familiar to anyone who has seen a polluted stream bubbling over a weir. This foam spills over into a plastic cup, where it collapses into a murky brown liquid which can be discarded. Motor-driven venturi protein skimmers are more expensive but even more powerful and efficient.

Counter-current protein skimmer

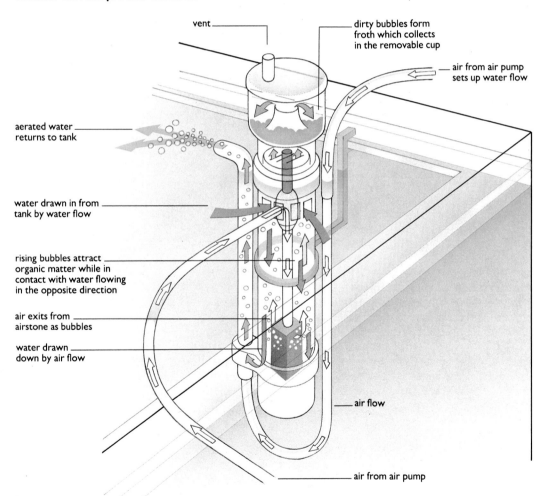

vent

dirty bubbles form froth which collects in the removable cup

air from air pump sets up water flow

aerated water returns to tank

water drawn in from tank by water flow

rising bubbles attract organic matter while in contact with water flowing in the opposite direction

air exits from airstone as bubbles

water drawn down by air flow

air flow

air from air pump

The efficency of protein skimmers is governed by their vertical height, the amount of air supplied to the unit and the water flow rate. The last two parameters determine the amount of contact between the stream of air bubbles and the water flowing through the unit. In practical terms, this means using the tallest unit that will fit in the tank, a strong air pump and a skimmer of the counter-current type. (In counter-current models the water flows against the current of the air bubbles, thus increasing the beneficial contact between the two.) Protein skimmers are rather clumsy looking units, but once in position they can be hidden behind rockwork and any servicing can be carried out from above, without removing the main body of the unit (see also page 90-1). The amount of waste removed by these units can be a revelation and their contribution to long-term success with invertebrates is second only in importance to good biological filtration.

Ozonizers

The popularity of ozonizers in the domestic aquarium has waned in recent years, but they are still widely used commercially. Here, a supply of dry air is pumped over a high voltage electrical discharge, where a proportion of the oxygen (O_2) in the air is converted to ozone (O_3). Ozone is chemically very reactive, a fierce oxidizing and disinfecting agent, and it was thought to be a useful background aid to disease prevention, since the gas kills bacteria and any other free-swimming micro-organisms that come into

Above: Large sea fans, such as this Caribbean specimen, are very efficient natural filters. Huge volumes of water pass across their feeding polyps as they extract fine particles of food and organisms.

direct contact with it. Their main commercial use, however, is to increase the efficiency of protein skimmers. By injecting ozone into the air supply to a skimmer, the efficiency rate of the skimmer is vastly improved. If you buy an ozonizer, do not feed the ozone directly into the tank, but couple it to a skimmer and regulate the generation rate so that you cannot detect the sharp tang of ozone around the tank. Always use ozone with extreme care; exposure to it can cause headaches, nausea and other effects.

Ultraviolet sterilizers

Ultraviolet light has long been known to have a strong sterilizing effect and UV lights have many industrial uses as bactericides. Units are available to the fishkeeper that pass precleaned water from a power filter close by an enclosed UV light. The light can kill algae spores and bacteria and may affect some small parasitic organisms. The efficiency of UV sterilizing units is governed by the internal cleanliness of the unit, the clarity and flow rate of the water going through it, and the intensity and precise wavelength of the UV light. If too much water is pumped through too small a unit, then many organisms will not be killed. If the correct size unit is balanced with the right water flow, UV units can be beneficial in reducing algae blooms. Their main use, however, is to reduce the incidence of disease among fishes kept in invertebrate systems, where chemical treatments are inapplicable or inadvisable. The output of a UV unit is rated by its wattage, 15 watts being the smallest available. For most purposes, a rating of 15 watts per 136 litres (30/36 gallons) tank capacity is adequate. The water in the aquarium should pass through the unit approximately twice per hour. The combination of a power filter and UV light can be very effective, if relatively expensive. Be sure to replace the UV tube about every six months for the best results.

Ultraviolet sterilizer

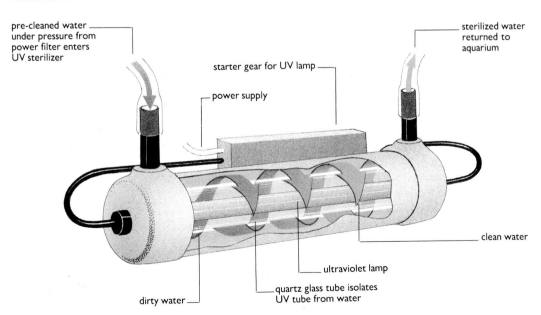

pre-cleaned water under pressure from power filter enters UV sterilizer

sterilized water returned to aquarium

starter gear for UV lamp

power supply

clean water

ultraviolet lamp

quartz glass tube isolates UV tube from water

dirty water

CREATING THE RIGHT WATER CONDITIONS

The welfare of any animal is ultimately dependent on the quality of the environment in which it lives. It follows, therefore, that one of the first vital steps to success with invertebrates is to use a high quality salt mix that closely matches the composition of natural sea water. Throughout the world, the chemical composition of sea water is remarkably similar. Although the total weight of salts in a given volume of water may vary slightly, particularly in comparatively land-locked waters such as the Red Sea, the proportions of each are largely the same.

Using salt mixes

In the early years of the marine hobby there were many 'sea salts' of varying qualities. Some were acceptable and others were certainly not! So much so, that many of the early books available gave detailed formulae so that the aquarist could buy the various chemicals required, blend them and hopefully produce something approaching sea water. Fortunately, those days are long gone and now there is a wide range of professionally produced salt mixes, the vast majority of which are perfectly suitable. As a general rule, it is a good idea to use the same salt mix as your supplier; although most salt mixes are virtually indistinguishable, there is no point in subjecting your livestock to any unnecessary environmental stress.

Generally speaking, any reputable salt mix can be dissolved in cold tap water to produce a satisfactory solution. One word of caution, however; in some areas, particularly of high population densities or where there is intensive arable farming, the tap water may contain nitrates at concentrations above those considered safe for some of the more delicate invertebrates. If there is any doubt as to the quality of your tap water, it is worth carrying out a simple nitrate test before mixing up new salt water. If it gives a reading much above 25mg/litre, either filter the tap water through a nitrate-removing resin before using it or find a safer source of fresh water.

Maintaining water quality

The main constituents of sea water are various sodium, magnesium, calcium and potassium salts. In addition, natural sea water contains small amounts of every known naturally occurring element. Examination of a bag of prepared salt mix will show that many other chemicals are also present in very small quantities. Certain of these so-called 'trace elements' are known to play a vital role in the well-being of both marine invertebrates and fishes, in that they are extracted from the surrounding water and help to sustain the animals' metabolic processes. In the wild, the concentration of trace elements is continually replenished through decomposition of dead animals and erosion of minerals from the land into the sea. Sub-oceanic volcanic activity may also be significant in this respect. Clearly none of these forces are at work in the domestic aquarium, yet trace elements are still being abstracted by the livestock and various algae in the tank.

Tropical fishkeepers will already appreciate the importance of making regular partial water changes as a means of removing

Composition of sea water

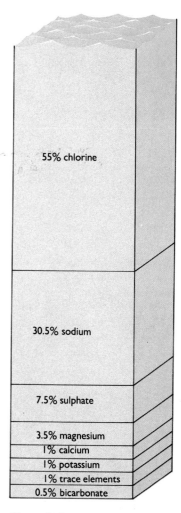

55% chlorine

30.5% sodium

7.5% sulphate

3.5% magnesium

1% calcium

1% potassium

1% trace elements

0.5% bicarbonate

Above: Sodium chloride, magnesium, calcium, potassium and sodium bicarbonate are all present in sea water. Trace elements occur in very small quantities.

Above: The warm, shallow waters of the Great Barrier Reef in Australia are home to huge variety of marine creatures. Synthetic sea water mixes make it possible to keep some of them in captivity, but it is important to monitor water quality carefully to maintain ideal conditions in the tank.

accumulated waste products that filters cannot eliminate. In a marine tank, regular water changes not only remove wastes, but also ensure that trace elements are replenished. Specifically formulated trace element supplements are available and are particularly useful in a heavily stocked aquarium or where water

Vitamin supplements can similarly be of great benefit. In the wild, much of the food taken by invertebrates is either alive – in the form of plants and planktonic animals – or freshly killed, and contains all their nutritional requirements. The marine hobbyist tends to rely heavily on frozen, dried or liquid foods. While most of these are vitamin enriched by the manufacturers, many vitamins have a comparatively short 'shelf-life'. Using a good vitamin supplement on a regular weekly basis will go some way to alleviating this shortfall. Be sure to keep liquid vitamin solutions in the refrigerator.

The pH of sea water

Another feature of sea water, compared to fresh water, is its high alkalinity. You can check this using a simple pH test kit, which should produce a reading of around 8.3. A reading below 8.1 is approaching a dangerously low level, while one above 8.5 (a rarity) may occasionally result in problems with excessive ammonia if the filter system is not fully biologically active, or if the aquarist tends to be heavy handed with his feeding. The point here is that a given amount of ammonia is more toxic at higher pH levels because a greater percentage of it is in the 'free' form (NH_3) rather than as ammonium ions (NH_4^+).

Synthetic sea water has a fairly high buffering capacity, i.e. it is able to maintain the correct pH level in the face of influences that might alter it. However, in the marine aquarium the animals' waste products are largely acidic and as a result the pH value of the water tends to fall (i.e. its alkalinity declines) over a period of time. The best way to counteract this tendency is by making regular partial water changes, say 10 percent every two weeks. Furthermore, many corals and molluscs extract the very chemicals that maintain pH levels in order to produce their shells and skeletons. Regularly using a pH buffering liquid, which by virtue of its formulation cannot take the pH value of the water into dangerously high levels, will maintain optimum pH levels, while at the same time ensuring there is sufficient calcium and magnesium, etc. available to the living corals, clams and other similar creatures.

Specific gravity

With the exception of areas such as the Red Sea, the salinity of sea water throughout the world is fairly constant and, within a particular small region, exceptionally stable. The salinity of water is a measure of the total amount of dissolved salts it contains; it is usually quoted as gm/litre (parts per thousand). Within the

The pH value of sea water

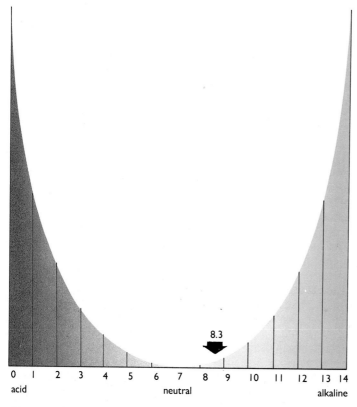

8.3

| 0 | 1 | 2 | 3 | 4 | 5 | 6 | 7 | 8 | 9 | 10 | 11 | 12 | 13 | 14 |

acid   neutral   alkaline

Above: The sea around coral reefs is chemically extremely stable. To recreate these conditions in the tank, it is vital to prevent major rapid changes to any of the water's parameters, such as the S.G.

Left: The pH value of water is a measure of its acidity or alkalinity. Sea water has a higher pH value than fresh water; in a marine invertebrate maintain the pH at about 8.3 for best results.

fishkeeping hobby, however, it is more usual to talk in terms of the specific gravity (S.G.) of water, since this is a good analogue, more easily tested and clearer to understand.

Specific gravity is simply a ratio of the weight of a water sample compared to the weight of an equal volume of distilled water at 4°C(39°F), which is assigned a specific gravity of 1.000. Since adding salts to water increases its weight as well as its salinity, the two scales are directly comparable. Thus, a salinity of 35gm/litre is equivalent to a specific gravity of 1.026 at 15°C(59°F). The traditional floating hydrometer used to measure specific gravity has been joined by easy to use and reliable swing-needle types.

Natural sea water generally has an S.G. of between 1.023 and 1.027, depending on the location. (It can be higher in some areas; in parts of the Red Sea, for example, it can reach 1.035.) Once properly acclimated, many invertebrates seem happy to accept a constant S.G. in a fairly wide range. For most purposes, a reading between 1.022 and 1.024 is acceptable. The key factor is to keep the S.G. as constant as possible. Using a hydrometer, take regular readings and ensure that they do not vary by more than one point (between 1.022 and 1.023, for example). Although this seems a minor shift, it is still many times that which occurs in even the most extreme circumstances in any one location in the wild.

In an established tank, the main cause of changes in specific gravity are evaporation and the addition of new stock. Over a period of time, the water level in the tank tends to drop as fresh water evaporates and thus the salt water becomes progressively more saline. Compensate for this by regularly adding a small quantity of *fresh* water. It is not good policy to wait until a large amount (more than 0.5 litre/approximately a pint, say) is needed, as the change in salinity will harm some delicate invertebrates and algae. Bear in mind that a small amount of salt will also be lost through emptying protein skimmers, cleaning power filters and by the accumulation of the crystalline deposits on cover glasses.

Two common, and dangerous, sequences of events can lead to a continual increase in the S.G. of the water in your tank. In the first scenario, you allow the water level in the tank to drop through evaporation and then decide to carry out a partial water change. Having drained out 10-20 percent of the water, you replace exactly the same amount with newly made up sea water, plus extra sea water to make up the evaporation loss. Since the reduced volume in the tank was more concentrated, this will cause an increase in S.G. Alternatively, the water level drops and you then add more stock (and its attendant sea water), thus filling the tank and also raising the S.G. Be sure to guard against these pitfalls.

In general, it is well worth noting how much water is lost by evaporation during the first week of a tank's operation, while the filters are becoming biologically active and there is no livestock in the aquarium. It is then an easy job to add, say, half a cup of fresh water to the tank on a daily basis. Nevertheless, you should always check the S.G. with an accurate hydrometer at least once a week. Tanks fitted with 'total management sytems' offer the ideal, if relatively expensive, solution since they incorporate self-acting topping-up devices that automatically replace evaporation loses from a reservoir of fresh water.

Specific gravity/salinity table

Specific gravity at 15°C(59°F)	Salinity (gm/litre or ppt)	Specific gravity at 25°C(77°F)
1.020	27.2	1.017
1.0203	27.6	1.018
1.021	28.5	1.019
1.022	29.8	1.020
1.023	31.1	1.021
1.024	32.4	1.022
1.025	33.7	1.023
1.026	35.0	1.0236
1.0264	35.5	1.024
1.027	36.3	1.025
1.028	37.6	1.026

Regular testing and maintenance: the key to success

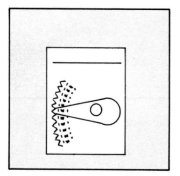

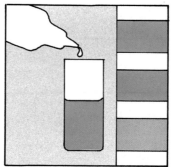

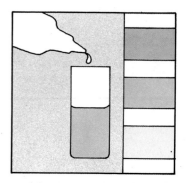

Check the specific gravity of the water at least once a week using an accurate hydrometer. Regularly replace water lost through evaporation. Keep the S.G. stable.

The pH value of sea water in good condition should measure 8.1-8.4. Check weekly; more frequent testing is advisable in heavily stocked invertebrate tanks.

Ammonia is highly toxic, both to fishes and invertebrates. Carry out an ammonia test immediately if any of the animals in the tank show signs of distress.

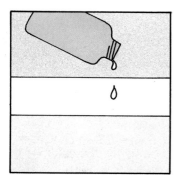

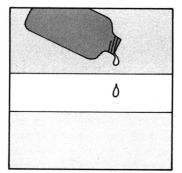

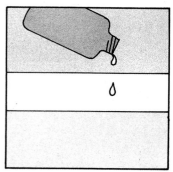

A good pH buffering solution or powder, added weekly, maintains the pH level at around 8.3 and ensures that calcium and magnesium are readily available in the tank.

Vital trace elements are extracted from the sea water by all marine animals to support their metabolic functions. Weekly supplements make up for any shortfall.

A weekly vitamin supplement replaces vitamins that are quickly lost from processed food. Particularly valuable in a tank where no live food is given.

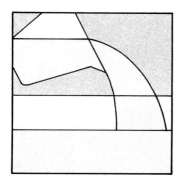

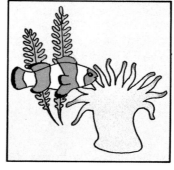

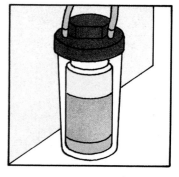

Partial water changes – about 10 percent per fortnight – are a valuable means of reducing many potentially harmful chemical end products that build up in the tank.

Check all stock and remove sick or dead animals at feeding times. Recognizing ailing stock and taking swift action will help to prevent serious problems developing.

Power filters need cleaning when visibly dirty or when the water flow slows dramatically. Change chemically active media at the first signs of exhaustion.

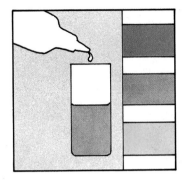

Nitrite is less toxic than ammonia. A weekly nitrite test is generally sufficient, but be sure to test newly established tanks at least every other day for the first three weeks.

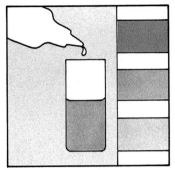

Nitrates, the end product of the nitrogen cycle, accumulate gradually and are particularly harmful to corals. A reading over 25ppm is cause for concern.

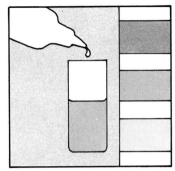

Copper is fatal to invertebrates. As a precaution, test for its presence in areas where fresh water is drawn through newly installed copper pipework.

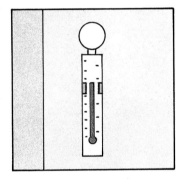

Check the thermometer twice daily, before you switch lights on and off. Temperatures between 24 and 26°C(75 and 79°F) will suit most marine invertebrates.

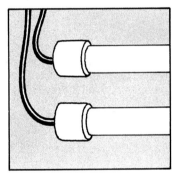

Most fluorescent tubes lose efficiency with age. Replace any that become dark near the end caps and protect them from being splashed by aquarium water.

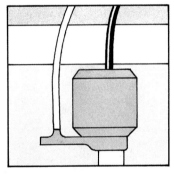

Efficient pumps are at the heart of the filter system. Check them at least once a day to ensure that they are working properly and have not become clogged up.

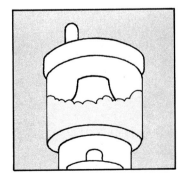

Empty the yellowish liquid that builds up in protein skimmer cups at regular intervals. This will depend on the stocking level and the amount of waste produced.

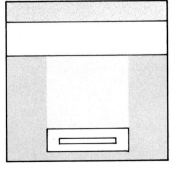

Algae on the front glass is harmless but unsightly. Remove it with a pad of coarse filter wool or an 'algae magnet'. Do not leave magnets permanently in the tank.

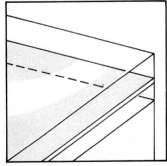

Clean cover glasses frequently to remove algae, salt crystals and dust, all of which can dramatically reduce the amount of light reaching animals in the aquarium.

SETTING UP

In this section, we consider the practical steps involved in setting up a marine invertebrate aquarium.

Decorative materials

Since many of the materials used to decorate the tank require careful and lengthy cleaning, let us first review the options available. Apart from the foundation rockwork, these decorative materials are installed once the tank is filled with water.

Calcareous stone was the only rock in regular use in the early days of the marine hobby. Although it is still perfectly acceptable, it is extremely heavy and does not have the many small crevices that many invertebrates like to colonize.

Corals were also widely used at one time for decoration, but are no longer as readily available and are rather expensive. Since the corals you buy from an aquatic store are the skeletal remains of living polyps, their bleached white appearance tends to look rather incongruous in a 'living-reef' situation.

Shells and barnacle clusters remain justifiably popular and provide valuable shelter for many small animals. However, unless you are certain that they are ready to use, you must clean all shells thoroughly before putting them in the tank. As a test, soak the shells in a bucket of fresh water for two or three days and then smell them. If there is no smell of decomposition they may go in the tank. If they do smell, place them in a solution of one cup of bleach and 4.5 litres (approximately 1 gallon) of water for a week, then boil them for half an hour and repeat the soak test.

Lava rock and tufa are the best rocks to use in the invertebrate aquarium. Tufa is a natural material, but lava rock is a man-made product. Do not confuse it with natural lava, which must never be used in marine tanks, since it is often rich in sulphur and metals that can cause severe pollution in sea water.

Lava rock is increasingly popular and rightly so. Although it is the most expensive stone on the market, it does have several advantages. It varies in colour between dark coffee and purple-brown and looks extremely attractive. It is also exceptionally light for its volume and easy to handle. Being fairly jagged, it stacks together well and is easily colonized by worms, sponges and algae.

Tufa is a calcareous rock, chemically very similar to coral sand. It is also very soft and easy to shape. When you first bring it home, it is often covered with a thick layer of aggregated dust, but when this is hosed off, the tufa often reveals some remarkably patterned surfaces. Many encrusting corals and algae quickly colonize tufa, and its initial stark white colour quickly tones down in the tank.

Sea fans and sea whips are sometimes available, but it is extremely important that they are first properly cleaned. Before cleaning, they have a chalky appearance and are often pink, yellow

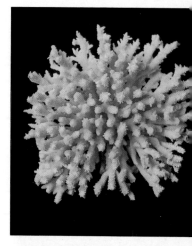

Top: *Acropora* skeletons are often available, and provide good cover for small shrimps and crabs.
Centre: Barnacle clusters are available at many aquatic stores.
Bottom: Many aquarists choose tufa rock to form the basis of their 'living reef' aquariums.

or buff in colour. When thoroughly washed in several changes of water, all that remains is a black wiry skeleton; unless they are in this condition do not use them.

All the above decorations can safely be added to the tank during the maturation period. This is not the case with living rock.

Living rock is the accepted name for pieces of calcareous rock taken from tropical seas, usually from a site close to the shoreline. When first removed from the sea, this rock is liberally clothed with various algae, sponges and miniature anemones, and usually houses a considerable population of small shrimps, crabs, worms and many other 'mini-beasts'. Some years ago, large quantities of living rock were shipped from Saudi Arabia and the short flight times to Europe meant that this rock usually arrived in very good condition and justified the name 'living rock'. As a result of various political constraints, this source of supply is no longer available, and most living rock now comes from the Caribbean or Singapore.

In view of its weight, living rock is shipped dry, i.e. not submerged in water, but packed in sealed plastic bags that retain a high level of humidity. As much of the rock comes from the tideline, many of the animals and plants living upon it are well able to cope with a period of several hours out of water, but invariably some of the more delicate organisms may succumb before reaching the dealer's tanks. Caribbean rock usually arrives in good condition, but it is always expensive and in limited supply. Rock from Singapore is cheaper, but because of the longer flight time much of the life which makes it so appealing is either dead or dying.

Below: A mixture of man-made lava rock, tufa rock and living rock provide the base for this 'living reef' aquarium. They have been colonized by various soft corals.

Some authorities recommend maturing marine tanks with living rock, but generally speaking it is not a wise practice, and newcomers to the hobby should never attempt it. The theory is that if you introduce approximately 0.45kg(1lb) of living rock per 0.45 litres (1 gallon) of water to a newly set up aquarium, you provide the tank with a good initial 'inoculation' of nitrifying bacteria, plus a thriving ecosystem of 'lower' animals and plants. For the next two weeks, carry out regular tests for ammonia and nitrite and only when the tests are clear, do you add any more expensive livestock.

This method can work well, but is more often a failure for a number of reasons. As we have seen, much of the life on the rock may be dead or dying when introduced to the tank and as it decomposes it produces unacceptably high levels of ammonia and nitrite. Although many of the animals on the rock are very hardy – a response to the fluctuating conditions in which they live – other species are unable to cope with such high levels of pollution. They die and further exacerbate the problem. In the worst cases, the tank's water may turn milky white and foul smelling, and a complete change of water and strip-down of the tank are the only remedy. In less extreme cases, the high levels of toxins can delay the maturation of the tank and so put back the day when stocking can begin. If you do give serious consideration to the living rock method of maturing a tank, make a point of smelling each piece of rock before buying it. Be sure to refuse rock that smells foul, or pieces with a white bubbly coating.

Good-quality living rock does, however, have an important part to play in establishing a thriving living reef system. When introduced to a bacterially mature system, it will encourage a

Right: It is an expensive exercise to use only living rock to the exclusion of all other dead or sterile forms of tank decoration, but the results can be spectacular.

Below: The top surface of good-quality living rock is often encrusted with polyp colonies, small sponges and featherduster worms. Small crabs, starfishes, shrimps and worms make their home within the rock and underneath you may find sea squirts, sponges and periwinkles.

population of micro-organisms, sponges and small shrimps, etc., that all play an important role in the ecology of the tank. Trying to produce a living reef that contains a good cross-section of invertebrate types without some living rock in the tank, is like gardening without earthworms, soil bacteria and beneficial fungi.

As a general rule, do not introduce more than 2.2-4.5kg(5-10lb) of living rock to the tank at any one time. When you buy the rock it will be packed in much the same way as it was when originally imported, i.e. sealed in plastic bags. Most of the water will have drained out of the rock, so be sure to tumble the rock over in the tank to release any trapped air before placing it in its final position. Burrowing molluscs, sponges and many worms – the very animals you want in the tank, and have paid for – will die if trapped in an air bubble. If they do, ammonia/nitrite pollution may result and the value of introducing the rock is diminished.

Remember, too, that all living rock has a top and a bottom, i.e. one area will have been exposed to light and another will have been shaded. If you put the algae-covered lit surface upside-down in the tank and expose the shade-loving animals to light, neither will survive in the aquarium.

Installing filters

If you opt for one of the very sophisticated total system filters, you will find that comprehensive details for assembling the system are provided with the equipment. In this book we are assuming that the aquarist intends using undergravel biological filtration with ancillary chemical/mechanical filtration.

Once the tank is in position, you can install the undergravel filters. Most manufacturers produce filters in a range of sizes that will fit most aquariums. It is important to cover as much of the base of the tank as possible, and small interlocking filter plates are very useful when dealing with an oddly shaped aquarium. Vertical uplift tubes are normally provided with undergravel plates. If you intend to use air pumps, fit them at this stage and ensure that the water level reaches approximately halfway up the outflow of the uplift. Uplifts may need cutting when used in conjunction with powerheads so that the top of the water pump is roughly level with the water surface. Most powerheads are water cooled and being submerged they avoid the risk of overheating. (See pages 65–66 for more information on air pumps and powerheads.)

Adding substrate

With the filter plates and pumps in place, the next step is to cover the plates with a layer of coral gravel mixed with a commercial buffering material, using approximately 2.25kg(5lb) of each per 0.1m^2 (approximately 1ft^2). Ensure that both materials are thoroughly rinsed before use to remove dust and small particles. Cover this coral gravel mixture with a perforated plastic mesh, known as a 'gravel tidy', and then add a further layer of well-rinsed coral sand to give a total substrate depth of about 7.5cm(3in). The

Above: After a few months, tufa and lava rock become encrusted with algae and look almost indistinguishable from living rock. These materials are widely available and easy to work with.

Right: Moderate lighting is often sufficient in a mixed fish and invertebrate aquarium, where only the more hardy invertebrates are being kept. Here anemones, bubble coral, polyps and various algae are in the company of a clown wrasse and colourful damselfishes.

mesh will help to prevent small grains of substrate from getting under the filter plate and possibly being sucked up into the powerhead, thus causing it to fail.

There is very little point in creating a filter bed more than about 7.5cm(3in) deep, unless you plan to keep those few animals that habitually burrow deeper than this. The theory that very deep filter beds will benefit filtration has largely been discredited. We now know that most of the beneficial biological activity in the aquarium takes place only in the first few centimetres of substrate that the water enters. Nitrifying bacteria demand very high levels of oxygen and little is available in the water once it has passed through 7.5cm(3in) of substrate. (In down-flow undergravel filtration systems, the top layers of substrate will be the most biologically active; in reverse-flow systems, the opposite applies, since the lower levels of substrate receive the well-oxygenated water first.)

If you wish, you can vary the level of the substrate by sloping it from the back of the tank down to the front.

Setting up the tank: a step-by-step guide to installing aquarium equipment

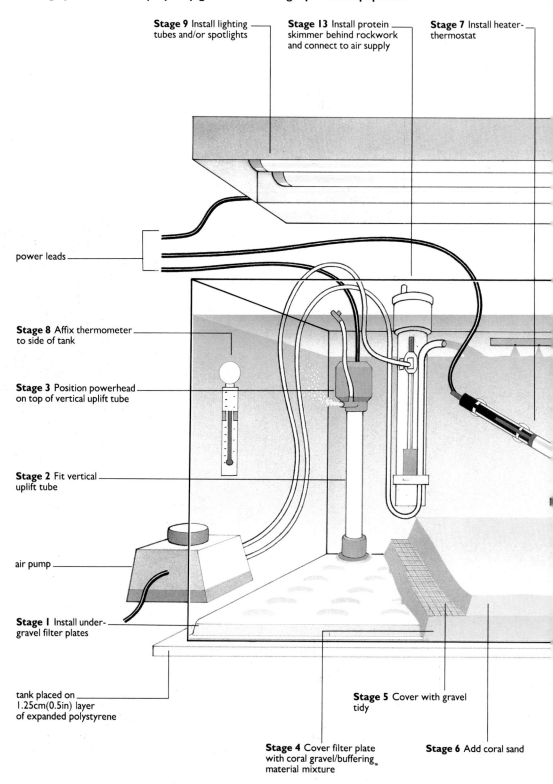

Stage 9 Install lighting tubes and/or spotlights

Stage 13 Install protein skimmer behind rockwork and connect to air supply

Stage 7 Install heater-thermostat

power leads

Stage 8 Affix thermometer to side of tank

Stage 3 Position powerhead on top of vertical uplift tube

Stage 2 Fit vertical uplift tube

air pump

Stage 1 Install under-gravel filter plates

tank placed on 1.25cm(0.5in) layer of expanded polystyrene

Stage 5 Cover with gravel tidy

Stage 4 Cover filter plate with coral gravel/buffering material mixture

Stage 6 Add coral sand

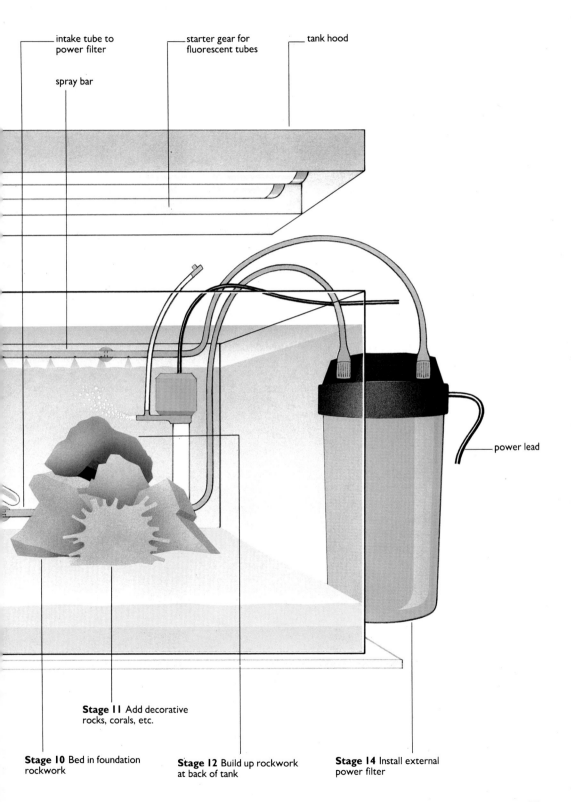

intake tube to
power filter

starter gear for
fluorescent tubes

tank hood

spray bar

power lead

Stage 11 Add decorative
rocks, corals, etc.

Stage 10 Bed in foundation
rockwork

Stage 12 Build up rockwork
at back of tank

Stage 14 Install external
power filter

Heating and lighting equipment

At this stage you can install the heater-thermostat unit in the tank. Choose a position that is out of the way, but still allows you to see the neon 'working' indicator light clearly. Almost all models are fully submersible, but are best not laid horizontally on the sand; mount them diagonally or almost vertically. Set the temperature at about 24°C(75°F) and then position a thermometer at the opposite end of the aquarium.

In view of the high water flow rates used in invertebrate tanks, there is very little risk of hot-spots developing. However, bear in mind that some invertebrates, particularly anemones, have a habit of moving slowly about the tank and in the process may adhere to a heater unit. When the heater next switches itself on, the anemone will be unable to move away quickly and may be severely burned. Such burns usually prove fatal, but can be easily avoided by the simple precaution of sliding a length of wide-bore, perforated, non-toxic plastic tubing over the heater.

Next install the lighting equipment. If you are using fluorescent tubes, make sure that there is no risk of the ballast units being splashed with sea water. The end caps that fit on the tubes are normally designed to be water resistant, but do bear in mind that sea water is an extremely good conductor of electricity and use splash trays or cover glasses. Fix one white tube towards the back of the lid and another at the front. Place one or more actinic blue tubes down the centre. (See pages 60–61 for more information on lighting equipment.)

Suspend mercury vapour and metal halide spotlights no less than about 23cm(9in) above the tank. If a supplementary actinic tube is used with these lights, house it within its own reflector; a length of half-round or 'U' section white guttering fits the bill perfectly.

Foundation rockwork

Now bed the foundation stones of the background rockwork into the gravel. Given the chance, many invertebrates will burrow under stones and it is much easier to anchor these securely when the tank is empty. Bear in mind the effect of refraction, which will make the tank look narrower once it is filled with water. When adding rocks to a dry tank, do not place them too near the front glass or it may look cramped once filled.

Filling the tank

Work out the volume of the tank and add enough of the dry sea salt mix to the tank to make up between 85 and 90 percent of the tank's volume. (To calculate the volume of an empty tank, multiply the length x width x depth – all in cm – and divide the result by 1000 to arrive at the volume in litres. Then multiply by 0.22 for Imperial gallons and 0.26 for US gallons.) For example, if the tank is nominally 181 litres (40/48 gallons) put in enough salt for 159 litres (35/42 gallons). Allow the cold tap to run to waste for a few minutes and then fill the tank to within about 5cm(2in) of the final water level. Alternatively, mix the salt and water in a non-metallic container, using a plastic or wooden spoon, before adding it to the tank. It is a good idea to fill the tank using a clean bucket of known volume so that you can establish the precise capacity of the tank.

Above: This magnificent collection of invertebrates demonstrates the stunning results that can be achieved with good filtration, lighting and water management.

Below: Using intense metal-halide lighting, the owner of this superb tank can stock a huge range of light-sensitive coelentrates and maintain them in peak condition.

When filling large tanks with a running hosepipe, first time how long it takes to fill a container of known volume and use this to calculate the volume of the tank. Thus, if a 22-litre (5/6-gallon) bucket takes a minute to fill and the tank takes 20 minutes, then the tank must contain 20 x 22 = 440 litres (100/120 gallons).

Powering up
Switch on the electrical supply and check that the pumps are working correctly, that water is being circulated through the undergravel filter, that the heater has started to warm the water and that the lighting is operational.

After 24 hours, the tank should be up to its working temperature and all the salt will have dissolved, leaving the water crystal clear. Use a hydrometer to check the specific gravity; if the reading is too low, add a little more salt, if it is too high, remove some water from the tank and replace it with cold tap water. When the specific gravity is correct, you can begin to add any non-living decoration in the form of rocks, shells, barnacle clusters, etc., to the tank.

Completing the decoration
To achieve a reef-wall appearance, build up plenty of rockwork in a honeycomb pattern with a few jutting platforms. Cover the back of the tank to within about 10cm(4in) of the final water level. A wide variety of sessile (non-moving) invertebrates will make their home in the many crevices and by adding the rocks the tank should be filled to the correct level – about 1.25cm(0.5in) from the top.

This type of rock arrangement is fine for corals, anemones, shrimps and, indeed, the vast majority of invertebrates but, as we have seen, it is impossible to satisfy the requirements of them all within one tank. A large lobster, for example, would probably knock over most of the rock and then hide behind the resulting pile!

Maturing the filter

Once the rockwork is in place, the process of maturing the filter system can begin. The gradual development of bacterial activity in the gravel generally takes place before the addition of any livestock. Since all animals' waste products are poisonous to them (see page 64), it is essential to produce a situation where the bacteria are 'ready and waiting' to go to work on these wastes before any valuable animals are added to the system. There are a number of ways of achieving this goal, the simplest of which is to use a maturing fluid or freeze-dried bacterial culture. Add these to the tank on a daily basis in accordance with the manufacturer's recommended dose until a nitrite test produces a reading of about 15mg/litre (ppm). Discontinue the dosage, but carry on with daily tests until the nitrite reading is nil. At this stage, there will be a moderate bacterial population and, providing the pH test kit and hydrometer both give satisfactory readings, the tank is ready to receive its first few animals.

Installing ancillary filtration

However, before adding any livestock, you should install ancillary filtration, i.e. the protein skimmer and external power filter. Place the protein skimmer in a rear corner of the tank, hidden behind rockwork. Here, any maintenance can be performed from above without disturbing the main body of the skimmer. Ensure that the slots in the top of the main column are at, or below, water level to

Above: With its heavy rockwork, this dimly lit aquarium is ideal for some of the larger and possibly clumsy invertebrates, such as lobsters, crabs and octopi. Little algae and very few corals would survive in this dull environment.

Right: A stunningly attractive aquarium such as this one is now within the scope of any aquarist prepared to invest in modern filtration and high-intensity lighting systems. Although expensive to set up, the results more than outweigh the costs involved.

allow water to enter the unit. When the two connections on the skimmer are attached to an air supply, a column of foam is generated within the plastic column, while water is pumped back to the tank through the skimmer return pipe. If both air supplies are valved, you can regulate the foam so that it rises slowly until it is above water level. The foam thickens and then collapses as it falls into the collection cup, leaving a brownish fluid that is discarded. You can drill a small hole into the cup, insert a length of thin plastic tube and direct this into a container that will only need emptying at infrequent intervals.

Position the external power filter either alongside or underneath the aquarium. Most cabinet aquariums have base units specifically designed to house such a filter. Load the canister from the bottom upwards with ceramic rings, followed by coarse mesh filter fibre, marine grade activated carbon and finally a thin layer of filter wool. Place the inlet tubing, with its strainer, in the tank about 2.5cm(1in) above the sand. A quick suck on the outflow pipe will start water siphoning down into the canister, which will begin to fill up. Water then rises up the outflow pipe to the level of water in the tank. Shake the canister gently to remove trapped air and prevent cavitation (i.e. the formation of gas bubbles in the water flow). Switch on the power filter and water will be drawn through the filter and then fed back into the tank. The system is now operational.

SELECTING AND MAINTAINING HEALTHY STOCK

All animals and plants are subject to a range of diseases, be they induced by bacteria, viruses or parasites. Unfortunately, because most invertebrates are of very little commercial value compared with other animals and because such a huge range of animals is involved, almost nothing has been discovered about diseases of invertebrates. Some good work has been carried out in the United States, where bivalve molluscs and crustaceans are farmed for the food market. Although several causative agents have been described, until recently there was little progress in the development of remedies suitable for the aquarium. Indeed, the commonest response to invertebrate disease was to destroy afflicted stock, sterilize the tank and restock. Now, the situation is beginning to improve. With increasing interest in marine fishkeeping generally, new products are beginning to appear on the market that do not

Above: Selecting good-quality, healthy livestock is obviously one of the major considerations when aiming for long-term success. Look for animals that are generally active, have no visible physical damage and are well expanded.

Left: Healthy shrimps are active and respond to any stimulus within the aquarium. Sponges should not display white or pale edges to the tissue and, where applicable, their siphons should be visible.

contain chemicals that will disrupt a biological filter system and are safe and effective in an invertebrate aquarium when used at the recommended strength.

Remember that the invertebrate keeper does have one advantage over the fishkeeper; not being involved in a monoculture (i.e. keeping just one species or family), the risk of losing entire stocks of animals to disease is negligible. Given good environmental conditions and an adequate source of food, vitamins and trace elements, most invertebrates appear very resistant to disease.

Many of the problems associated with invertebrates are either environmental or the result of incompatibility with tankmates. With care and consideration on the part of the aquarist, many of these can be quite easily resolved. However, bacterial infections are a different matter. The majority of aquarium bactericides available commercially are primarily designed for treating freshwater or marine fishes. Unless you are sure that they are safe in an invertebrate aquarium, refrain from using any chemical medication because the potential risks are usually far greater than the possible advantages. However, the appropriate use of UV sterilization and ozone as methods of reducing the background bacterial count can prove very valuable, with a UV system being preferable if a choice must be made between the two.

It has often been said that prevention is better than cure, and it is certainly true that you can do a great deal to reduce the chances of disease. Start by selecting only good-quality specimens, with no signs of physical damage that could lead to bacterial infections or, in the case of starfish, for example, to the rapid and complete collapse of the animal. The following guidelines will show you what to look for when selecting and maintaining stock.

SPONGES

● **Selecting**: Ensure that the base is attached to a small piece of rock and not buried in the substrate. There should be no discoloured patches and no overgrowth of film or hair algae.

● **Maintaining**: Many aquarists encounter problems with sponges, yet these are intrinsically among the most resilient of animals. Much of the difficulty lies in not appreciating the natural lifestyle of these animals. They are normally found in dimly lit areas, where they do not become coated with algae, and where there is a moderately strong water flow to bring them oxygen and food and take away waste products. Few species can withstand exposure to the air, as this is so easily trapped within the body, thus preventing the various metabolic processes.

Sponges are food for many animals and they are easily damaged by contact with the stinging cells of coelenterates. Damaged areas are very pale in colour and commonly occur at the tips of branches and at the base, where the animal has been removed from the substrate. Place sponges on rocks – not on the sand substrate of the aquarium – to allow good water movement around them. In good conditions, sponges grow rapidly, anchoring themselves in position in the process.

COELENTERATES

● **Selecting**: Anemones should have expanded tentacles and the central mouth should be closed. A healthy anemone will show a reflex reaction when disturbed, either by withdrawing its tentacles and rapidly shrinking in size, or by becoming very tense and firm. In *Cerianthus* species, the anemone should withdraw into its tube. Look for evidence of damage. Animals may have heater burns, or the flesh – particularly the foot – may have been torn during capture.

Polyp colonies and mushroom polyps should be expanded, but contract or stiffen when disturbed. There should be no visible damage, or any smell of decomposition when the colony is brought to the surface of the water.

Corals should be firm and plump and intact. Pay particular attention to the base; leather corals are often damaged during collection and rapid decomposition spreads quickly, often destroying the animal within a few days. *Sarcophyton* species regularly shed a skinlike mucus, but this is of no concern.

When selecting soft corals, the main precaution is to check for areas of almost powdery decomposition. Regularly siphon away any debris that collects in the dishlike structure of many soft corals.

The thin, colourful flesh that covers the horny skeletons of sea whips and sea fans is easily damaged. In perfect conditions, these lesions may heal and the animal recovers but, more commonly, the bald areas spread and the animal eventually dies. The most common points of damage are the extreme tips of the branches and the base, where the animal was fixed to the seabed. The best specimens are those that remain attached to a small piece of rock.

Stony corals, such as *Goniopora* and *Euphyllia* species, should be expanded and erect. The coral head, as it is called, should have no bald patches, or areas of grey, bubbly decomposition.

● **Maintaining**: Although an ailing specimen will generally die quite rapidly, healthy anemones are some of the hardiest invertebrates. However, air or gases resulting from decomposition show up as pale areas within the body which, because of the buoyancy involved, seems to stretch upwards. Remove any anemone with these symptoms from the tank, as it will decompose very rapidly and can cause a major pollution problem. From time to time, most anemones will contract quite dramatically to void the contaminated fluid that accumulates within the body. A string of thick brown mucus may also appear. This is quite normal and the animal should reinflate within a few hours.

Many aquarists do not appreciate that interspecific fighting among corals and anemones is a common cause of death among these animals. To allow for this, leave a gap of at least the diameter of a given coral between that one and the next. This also allows for the fact that many corals and anemones do not expand to their fullest extent until after dark. Anemones and corals may also shed undischarged nematocysts (see page 14) and can therefore sting each other at a distance. An efficient protein skimmer or power filter will reduce this risk, but it is not a good idea to keep anemones of different families within the same tank.

Above: Anemones are easily damaged during collection. Inspect the basal disc for tears, as any specimens so damaged almost invariably die within a few days. The tentacles should be expanded and usually retract when touched.

Below: Damage to gorgonian corals is common. In the wild such specimens may recover, but this is generally the start of a slow decline for sea whips in tanks.

Below: This *Goniopora* coral has several small 'bald' areas and is not a suitable specimen for inclusion in the tank. In healthy corals, the tissue should cover all the skeleton, except the base, and all the polyps should be well extended.

With very few exceptions, all corals and anemones appreciate a strong flow of water. If the flow is too slow, a 'skin' of almost still water forms around the animal, which is then unable either to catch food or rid itself of its own metabolic waste products.

Stony corals require very good water and lighting conditions. Provide a high pH level and a good reserve of calcium in the water. When lighting the aquarium, bear in mind that corals and anemones can be loosely divided into two groups. Those within the colour range of purple, through red to orange and yellow are lacking in zooxanthellae and generally do not require intense lighting. In contrast, those with beige, brown, green or blue colouring require strong lighting of the correct spectrum if the algae are to function correctly. Many of the problems encountered with these species can be traced back to poor environmental conditions. The corals fail to expand and feed, and the flesh gradually shrinks until only the rocky skeleton remains.

Another common problem, particularly with *Goniopora* species, is physical damage leading to bacterial infection. If their soft tissue is punctured, possibly by a tumble within the tank, the action of predators or through inadvertent contact with sharp-footed tankmates, then bacterial infection often follows. This commonly takes the form of a greyish brown disc of slime that rapidly enlarges, while the remaining flesh lifts away from the underlying skeleton. Gently siphoning the decomposing tissue away may slow the process, but the long-term prognosis is generally poor.

Corals should react in a similar way to sea anemones when handled. Once again, do not remove them from the water unless absolutely necessary and never expose them while expanded, as the weight of water inside them will almost invariably cause the delicate flesh to tear and allow bacterial infections to set in.

PLATYHELMINTHES

● **Selecting**: Flatworms should be active, with no tears in the body, and should seek shelter when disturbed.

● **Maintaining**: Do not house these delicate animals with potential predators, such as crabs, lobsters and large shrimps. Very little is known about the diet of these creatures, but they may accept finely chopped shellfish and similarly sized particles. Scavenging species may benefit from a good background growth of algae.

ANNELIDS

● **Selecting**: The main species of interest to aquarists are the various tubeworms that live within parchmentlike or stony tubes. These should show evenly shaped heads of feathered tentacles and the worms should withdraw rapidly into their tubes when disturbed. Avoid specimens that hang limply from their tubes or do not retract when touched.

● **Maintaining**: These animals do not demand the bright lighting required by many other invertebrate species, but they do enjoy a moderate to good flow of water. In the wild, they are efficient filters of both zooplankton and phytoplankton (vegetable matter). In the aquarium, this dietary requirement is easily met with liquid foods.

Above: This featherduster worm, *Sabellastarte* sp., is very close to death. These worms occasionally shed the feathered head, but the body should remain in the tube.

Below: Crustaceans often lose a claw during transit, or in a conflict. Missing limbs are generally replaced, provided the animal can continue to feed unmolested.

CRUSTACEANS

● **Selecting**: As a general guide to the health of crustaceans, check the mouthparts. All crustaceans have a complicated system of mouthparts that are forever in motion. If there is no sign of movement around the mouth, and the animal does not respond to disturbance within the tank, then it is ailing. Crabs, shrimps and lobsters normally make determined efforts to avoid capture – hermit crabs, for example, should retract rapidly into their shells. Do not buy lethargic animals.

Crustaceans should have their full complement of limbs, but provided the animal is not subject to predation and can still feed, the loss of one or more legs or claws is neither unusual nor cause for concern, because these will be replaced at the next change of exoskeleton. Deformed or broken antennae are of little significance. Check shrimps and crabs for fungal growths.

● **Maintaining**: If a shrimp or crab is injured in the tank to the extent that it may be at risk from one of its tankmates, isolate it in a perforated plastic container within the tank (floating breeding traps are ideal) until the missing limbs are replaced. Keep a check on the pH level of the aquarium water; if it is too low, the new shell may not harden properly and the animal will become deformed.

The internal fungal infections, to which shrimps and crabs are susceptible, typically show up as white patches within the body or

Isolating injured invertebrates

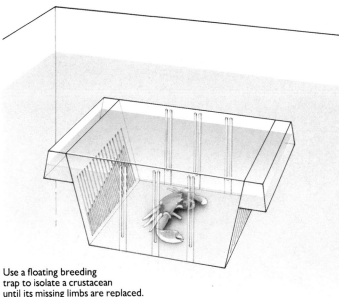

Use a floating breeding
trap to isolate a crustacean
until its missing limbs are replaced.

as discolorations on the surface. Some crabs suffer from a fungus
that invades the entire body and is visible externally as a tufty
growth on the underside of the carapace (the main 'body shell'). At
the moment, there is no readily available medication for this
problem. Generally, it is best to remove infected animals in the hope
of preventing transmission to related species.

MOLLUSCS

● **Selecting**: In clams, scallops and mussels, the mantle should be
expanded and the breathing siphons extended. These bivalves
should close quickly if disturbed; clams with mantles falling
inwards, away from the shell, are dead or dying. Univalves, such as
cowries, should withdraw quickly into their shells if touched. Check
the shells of bivalves and univalves for major physical damage.

The shell-less snails, or sea slugs, should be active and, where
applicable, the gill tufts on the dorsal surface should withdraw or
contract if the animal is touched.

Octopus, squid, and nautilus, etc., all belong to the subgroup
Cephalopoda. They are mobile animals that should actively try to
evade capture. Tentacles should be firm and mobile, with no major
physical unhealed damage. The complete loss of an arm, provided
the site has healed, is not critical.

● **Maintaining**: Most of the problems encountered with molluscs
tend to be associated with their environment and, of course, many
species are predated upon by other tank inhabitants. Furthermore,
most molluscs are continually extracting calcium from the water to
enlarge their shells and any shortfall in this essential mineral can
cause severe problems, particularly among the bivalves.

Octopi and their close relatives require perfect water conditions and very slow, careful acclimation to a new environment, preferably in semi-darkness. They are also very sensitive, nervous animals and losses can often be ascribed to shock. When disturbed, most species are able to eject sepia, an inklike substance that provides a smokescreen to cover their escape from predators in the wild. In captivity, 'inking' can cause a pollution problem, particularly in a small aquarium.

Above: Only consider keeping an octopus or its relatives if you can provide perfect water conditions. These shy animals also need hiding places or they suffer from stress.

ECHINODERMS

● **Selecting**: Echinoderms generally deteriorate very rapidly if infected or damaged, so you can be fairly confident of selecting a healthy specimen.

Starfishes should have plump bodies and be firm to the touch, with their tube feet expanded. Any flaccidity in the arms may be a sign of failure of the internal water vascular system, which is almost invariably fatal. Introducing specimens too rapidly into a new tank may also produce this symptom, with a similar end result. Damaged arms often show white eruptions; reject specimens in this condition. It is worth checking the underside of the arms for damage and parasitic molluscs. Healed scars can be ignored because, in the wild, starfishes show remarkable regenerative abilities, but these are largely lost in captivity. Obvious mechanical damage is clearly a serious disadvantage.

Sea urchins are quite easy to choose. The spines – and the tube feet among the spines – should be erect and mobile; indeed many species will actively point their spines towards questing fingers. If the spines are depressed, giving a 'thatched' appearance, the urchin is almost

certainly dying. You may notice small bald patches of missing spines, but as long as the tube feet are still active, the animal will generally regrow new spines and recover. However, if both spines and tube feet are absent, reject the urchin.

Sea cucumbers should be plump and, if the feeding tentacles are expanded, these should be largely intact and should retract rapidly when the animal is disturbed.

Crinoids, or feather stars, are very brittle animals and easily damaged. The loss of one or two arms may not be too harmful; in good conditions, they will slowly regrow, but any further damage is not acceptable. Be sure to check that the small, clawlike, gripping appendages are present on the underside of the animal.

● **Maintaining**: Some species of sea urchin seem particularly prone to getting air trapped within the body, which invariably proves fatal. In view of this, never remove urchins from the water when transferring them from tank to tank. Place them in a small, clean, plastic tub or thick plastic bag and allow them to become slowly acclimated to the conditions in the new tank. All echinoderms require this period of gradual readjustment.

Sea cucumbers are very resilient animals, but one fairly common problem is damage caused by heater burns. These very slow-moving animals occasionally settle on heater units and when the thermostat switches on, the cucumber is rapidly and severely burnt. Avoid this risk by hiding the heater behind rockwork, protecting it with a perforated plastic tube or using an outside heat source. The feathery tentacles of the more commonly kept species are often targets for predators. As the tentacles are used to trap food, their loss is obviously serious, but they usually regenerate over a period of several weeks, provided the predator is removed.

If you see threads protruding from the anus of a sea cucumber, the animal has probably been incorrectly acclimated and has voided part, if not all, of its intestines. In the wild, this behaviour deters predators and the cucumber normally regenerates its internal organs. In captivity, such evisceration is usually followed by death.

The commonest problem in crinoids is major limb loss, with numerous arms being broken. The arms are very flexible in a vertical plane, but have very little lateral fexibility. If you place the animal in a fierce current, such as is produced by some of the submersible powerhead water pumps, many arms may snap off in a short period. Remember that predatory or heavy, clumsy animals, such as cowries, can cause considerable damage to brittle crinoids.

CHORDATES

● **Selecting**: While most commonly introduced by accident to the tank, a few of the larger, more colourful species of sea squirts occasionally appear on the market. Their breathing and feeding siphons should be open, but quickly close when the animal is disturbed. Check the base and sides carefully for tears.

● **Maintaining**: These animals require the same care in captivity as featherduster worms (see page 96).

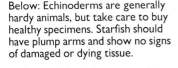

Below: Echinoderms are generally hardy animals, but take care to buy healthy specimens. Starfish should have plump arms and show no signs of damaged or dying tissue.

Diseases of fishes in the invertebrate aquarium

It is clear that including fishes in the invertebrate aquarium adds another dimension to the underwater scene. But you must be clear where your priorities lie. Is this primarily a fish tank or an invertebrate tank? This basic decision assumes a greater significance when you are faced with the problem of coping with fish disease in the invertebrate aquarium. It is almost certain that at some time you will find yourself facing an outbreak of marine white spot disease (caused by *Cryptocaryon*) or coral fish disease (caused by *Amyloodinium*). The most effective medication for treating both these protozoan infections is based on copper sulphate. Copper sulphate kills invertebrates. The dilemma is clear.

If you are primarily interested in fishes, then you can remove the invertebrates and treat the fishes with copper sulphate. If the aquarium is intended primarily for invertebrates, and you cannot remove the fish to a separate tank, then they must survive as best they can. (Indeed, in a well-balanced environment many fish will recover on their own.)

In the worst of all possible worlds, you may find yourself with a valuable stock of ailing fish housed with a cherished collection of invertebrates and no hospital or quarantine tank ready for use. You may allow the first signs of disease to progress in the fish for three or four days until a heavy infestation develops, before deciding, almost invariably, to save them in preference to the invertebrates. You dispose of the invertebrates, as these are of no use to anyone else – having come from an infected aquarium – and treat the tank with a copper-based medication. Unfortunately, at this stage, it is often too late to save the fish and you have lost everything.

This is clearly an unsatisfactory state of affairs, yet until recently there was no effective but safe alternative to copper sulphate (see page 92). Even when using those medications described as safe in the invertebrate tank, take care that the overall ecology of the tank is not upset. Although UV units can help with disease prevention, (see page 73) it is in your best interests to ensure that any fish added to the invertebrate tank are in peak condition. All responsible dealers, both wholesale and retail, do their utmost to ensure that they offer healthy stock for sale, but no animal can be guaranteed free of disease organisms. Neither can you be sure that your tank does not contain them.

Many fish diseases are termed 'stress-induced', i.e. the strain of being moved, coping with a change of water, or having to compete for territory in the new tank reduces the fishes' resistance and allows a previously benign organism to develop into an overt disease syndrome. It may help to quarantine new stock for two or three weeks in a tank where a full spectrum of medications can be used. Always ensure that the water quality in the quarantine tank is up to the same standard as that in the main aquarium. However, even then the final move to the show tank can be stressful.

Should a small fish die in the tank, it will normally be disposed of quite rapidly by scavenging animals. Be sure to remove any large corpses to prevent ammonia and nitrite surges, but take great care when moving rockwork in the process; it is all too easy to damage corals, crush shrimps and crabs, and puncture starfishes in the process. In a very large aquarium, it may be totally impractical to

The living reef aquarium ecosystem

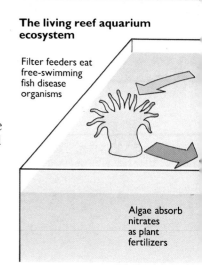

Filter feeders eat free-swimming fish disease organisms

Algae absorb nitrates as plant fertilizers

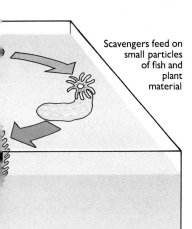

Scavengers feed on small particles of fish and plant material

move substantial amounts of rock in the search for a missing fish. In this case, monitor ammonia and nitrite levels for a week after the fish has died or disappeared and counter any increase in these levels with a series of small water changes. By the end of a week, scavengers should have disposed of the remains.

However, there is a more positive side to the picture. For example, in the well-balanced ecosystem of a good living-reef aquarium, filter-feeding invertebrates feed on free-swimming fish disease organisms, micro-scavengers tidy away small particles of food and macro-algae reduce the build-up of nitrates within the tank. Such a tank is generally freer of disease than a typical fish-only aquarium. Water quality in an invertebrate aquarium is generally better than in a fish-only tank, and the interest and colour of the invertebrates may reduce the temptation to overstock with fishes. Maintaining good water quality, low fish-stocking densities and avoiding interspecies aggression are the major factors leading to good fish health in invertebrate aquariums.

Symptoms	Causes	Treatment
Sprinkling of very small yellowish white spots all over the body. Fishes rub themselves against rocks and stones and may show rapid gill movements	Coral fish disease caused by the single-celled parasite *Amyloodinium ocellatum*	Remove fish to hospital tank and treat with recommended course of copper medication. Try one or more freshwater dips of 2–10 minutes. Quarantine new fish and treat with copper medication as a precautionary measure
Small white cysts on skin, fins and gills. Cysts may join together, forming irregular white patches	Marine white spot disease caused by the single-celled parasite *Cryptocaryon irritans*	Remove fish to hospital tank. Treat with proprietary remedy. Follow maker's instructions regarding length of treatment. One or more freshwater dips of 2–10 minutes may help
Film of greyish white mucus producing a noticeable sliminess of the skin. Swollen gills, rapid gill movements	Several likely causes, including single-celled parasite *Brooklynella*, marine white spot (*Cryptocaryon*), fluke (*Benedina*), and flatworm infection	Use copper-based and anti-parasite remedies in hospital tank. One or more freshwater dips of 2–10 minutes may help. Check water quality and diet
Cauliflower-like growths on skin and fins. Growths may also develop on internal organs	Rough growths developing from white cysts may be caused by the viral infection lymphocystis. Firm swellings may be tumours	Growths do not always cause the fish distress. Surgical removal is possible but not always successful. Improve general tank conditions. Painlessly destroy severely affected fish
Darkening of coloration, development of ulcers, lesions and loss of appetite. Vent and fin edges become red and inflamed	Ulcer disease, a bacterial infection caused by *Vibrio anguillarium*	Isolate affected fish. Treat with copper-based remedy and antibiotics (seek veterinary advice if necessary). Improve poor aquarium conditions
Loss of balance. Fish unable to maintain position in tank	Swimbladder inflamed and malfunctioning. Chilling is a possible cause	Bathe fish in water about 5°C(9°F) warmer than stock tank, raising temperature gradually. Antibiotics may help; seek veterinary advice

FEEDING

The whole question of what, when and how to feed invertebrates causes confusion among beginners and experienced hobbyists alike. A number of useful 'invertebrate foods' are available commercially, both frozen and in liquid form. Certain brands are very good, and the manufacturers are continually improving their products, but none should be considered as a complete diet for all invertebrates. Their main value is as a substitute planktonic diet for filter feeders, such as corals, featherduster worms, some molluscs and sea cucumbers. You have only to consider the varying roles that invertebrates fill in the ecology of the seas to realize that no single type of food can meet the dietary requirements of all species of invertebrates.

Invertebrates appear at every level of the classic food pyramid, from primary food generators in the form of corals, micro-crustaceans, etc., to top predators such as octopus, squid and some large crustaceans. Clearly, a suspension of fine particulate food will not satisfy an octopus, while the frozen shrimps that they appreciate are way beyond the capabilities of some of the attractive filter feeding worms that extract very fine particles of food from the water around them.

For practical purposes, it is convenient to divide invertebrates into three feeding groups: the hunters; the opportunists and scavengers; and the filter feeders and self-sustainers.

Hunters

A number of invertebrate groups are active predators, which means that they are capable of catching active prey animals. While most of these are of little interest if you plan to keep a 'living reef' aquarium, you may see them for sale and you may even introduce them into the aquarium by accident, as we see below.

The cephalopods – octopus, squid, cuttlefish and nautilus species – are becoming increasingly popular, but only the octopus species are regularly available, although generally quite expensive. All are fascinating creatures that quickly overcome their initial shyness when introduced into the aquarium. Unfortunately, they will eat almost any fish, shrimps, crabs or lobsters with which they are housed. Indeed, these animals should be fed a diet of defrosted whole shrimps, prawns and small fishes. The amount will vary with the size of the cephalopod, but an octopus with a 30cm(12in) span would need 2 or 3 shrimps, or a 5-6cm(2-2.4in) fish daily.

Equally predatory mantis shrimps and swimming crabs are often introduced accidentally with newly imported living rock. Both are highly active predators known to kill and eat shrimps and quite sizeable fishes. Never introduce either of them into an aquarium containing a mixed collection of animals, as expensive losses will undoubtedly soon occur.

Opportunists and scavengers

Many of the most attractive, desirable and easily maintained invertebrates are included within this grouping. The vast majority of shrimps, crabs and starfishes (and some molluscs) will eat

Above: This large *Choriaster* starfish is able to eat quite large food items, thanks to its extensible stomach. It will also eat many sessile invertebrates. Providing a suitable diet is not difficult, but chose tankmates with care.

Right: The beautiful orchid shrimp, *Hymenocera* sp., is one of the most specialized feeders regularly on sale. It feeds almost exclusively on starfishes, as shown here.

Below: Unlike most of the similar looking cowries, the egg cowrie has a limited diet and is particularly fond of the *Sarcophyton* leather corals, which it can devour at an alarming rate.

almost anything edible that comes their way. Their needs are easily met by feeding them with small pieces of fish, or various shellfish. These animals can subsist on surprisingly small amounts of food and you must take care that the tank does not become polluted with excess uneaten food. A certain amount of trial and error regarding quantity and frequency of feeding will be necessary but, as a guide, offer as much as the animal can eat in 10 minutes every other day.

Bear in mind that with their wide and catholic tastes, many species will quite happily eat the other sessile (non-moving) invertebrates that you hope to culture. Some of the starfish are well-known offenders in this respect – for example, the colourful red and white *Fromia* eats sponges. As a general rule, those species with large knobs on the dorsal (top) surface will eat almost anything upon which they can settle. Some of the small, brightly coloured dwarf lobsters may also prove somewhat destructive.

Some invertebrates, such as several molluscs (*Sarcoglossan* sea slugs, for example) and sea urchins will browse on algae in the tank. You can feed these largely herbivorous species with other vegetable matter, such as lettuce, spinach (in moderation) and seaweed preparations from health food stores.

Filter feeders and self sustainers

As their name suggests, filter feeders have evolved feeding mechanisms that enable them to trap small particles of detritus and planktonic organisms. Many worms, sponges and corals fall within this group. Anemones can also be considered as filter feeders, but they trap much larger particles of food than other species. Offer them one or two pieces of fish, shrimp, squid etc., once or twice a week. Gently push the food into the tentacles; do not force it into the mouth and remove any pieces not enfolded by the tentacles.

The self sustainers include many of the shallow water coral species and some sea anemones. All of them house various species of algae within their tissue, known collectively as zooxanthellae (see page 16). As we have seen, zooxanthellae rely heavily on lighting of the correct spectrum in order to survive. In good conditions, where the host animals are given sufficiently intense lighting, the algae will prosper and the animals need very little supplementary feeding. However, under poor lighting conditions the algae die, followed fairly shortly by the host animal. If necessary, add a few drops of a good liquid feed per animal every other day. Avoid over-feeding, however, otherwise ammonia and nitrite pollution may ensue.

Types of food

Most of the supplementary feeding requirements of these and other invertebrates can be met quite easily with readily available prepared foods. Most of the frozen foods are useful, but the dried foods familiar to freshwater fishkeepers have little place in the invertebrate aquarium. The only possible exception is the finely powdered fry food, sold for newly hatched fish larvae, which is acceptable to some filter feeders.

Liquid feeds are very popular and a convenient food source for filter feeders, but a freshly prepared solution is preferable to one that may have been lying on a shop shelf for weeks or even months. To prepare a solution, simply force a piece of shrimp, fish, or mussel, etc., through a fine meshed net and allow the resultant liquid to drip into the tank. You can prepare a larger supply of food by liquidizing a 10mm(0.5in) cube of fish, mussel, squid or shrimp, straining the liquid through a net to remove any larger particles, and then diluting it with 0.28 litre(0.5 pint) of sea water. Freeze this solution in ice-cube trays ready for use as needed.

It is generally safer to offer only gamma-irradiated sterilized frozen foods. Although invertebrates appear less susceptible than fishes to disease introduced via non-sterilized foods, it is unwise to take any avoidable risks.

Other valuable foods for filter feeding invertebrates are the newly hatched nauplii stages of brineshrimps, which are easily produced from eggs available from aquatic dealers, and the rotifers and algae used by marine fishkeepers in their efforts to breed and rear stock. These latter foods are fairly simple to produce if you follow the detailed instructions provided by suppliers of the initial cultures. Unfortunately, however, the space and time required for the process is more than most hobbyists are prepared to give.

The table opposite provides a summary of suitable foods for the different invertebrate groups and a guide to frequency of feeding.

Above: Sea anemones are easy to feed. Not only do they produce much of their nutritional needs via the zooxanthellae in their tissues, but they also accept quite large pieces of fish, shrimps or shell meat.

Feeding table

Hunters	Opportunists and scavengers	Filter feeders and self sustainers
Large crabs, lobsters	Small crabs, shrimps	Sponges
Large univalve molluscs, e.g. Queen conch	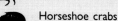 Horseshoe crabs	Sea pens, soft corals, sea whips, sea fans, anemones*, jellyfish, hard corals
Octopus, squid, cuttlefish, nautilus	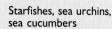 Cowries, sea slugs	Flatworms
	Starfishes, sea urchins, sea cucumbers	Fanworms, tubeworms, featherduster worms
		Barnacles
		Clams, scallops, oysters
		Crinoids
		Sea squirts

Suitable foods (Hunters)

Small whole fish, shellfish meat, defrosted whole shrimps, prawns. Flaked food tucked into fish meat to supply essential trace elements

Quantities and frequency of feeding

Feed once a day. Remove uneaten food after one hour. Quantities depend on the size of the animal, e.g. 12in diameter octopus needs 2–3 large shrimps, or one 5–6cm (2–2.4in) fish

Suitable foods (Opportunists and scavengers)

Small pieces of fish, shellfish meat, shredded prawn, mysis shrimp, brineshrimp, cockles, mussels. Algae
Good quality flake food

Quantities and frequency of feeding

Feed once a day at most. remove uneaten food after one hour. These animals require little food and may subsist on the leavings from any fish in the tank

*Anemones are classified here as filter feeders, but will accept large pieces of food once or twice a week

Suitable foods (Filter feeders)

Liquidized fish, shellfish meat or mussel blended with salt water from the aquarium. Newly hatched brineshrimp nauplii. Algae, pulverised or as a suspension.
Prepared plankton food
Liquid feed
Vitamin supplement

Quantities and frequency of feeding

Feed once a day at most. Filter feeders are better underfed than overfed. Under intense lighting many corals and clams need little if any supplementary food. One drop per day per animal of a good liquid feed is sufficient

Left: The beautiful, but rarely seen, nautilus is an avid predator, eating small fish and crustaceans with gusto. Overfeeding it will put a strain on filtration systems.

PART THREE

INVERTEBRATES
FOR THE AQUARIUM

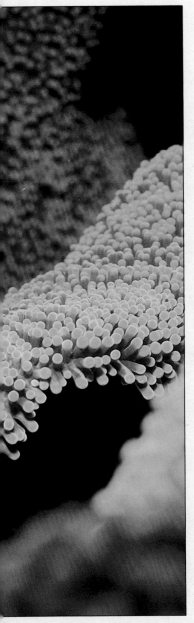

Fortunately for the aquarist, few tropical marine invertebrates are threatened with extinction, but there is concern that some may be declining in number. For example, the queen conch and the spiny lobster have traditionally been important food sources for coastal people, but are increasingly becoming food for the rich tourist, and some popular shells have been over-collected for the ornamental shell trade. Collecting corals is a serious threat to many coral reefs because removing the corals also damages the basic framework of the reef. If the reef is suffering from pollution, or intensive recreational use, the combined effects can be disastrous. Sometimes the methods used to collect marine invertebrates can damage the habitat, making it more difficult for the native population to recover. Rocks and coral blocks may be left overturned and coral reefs may be damaged by boats and anchors.

On a more positive note, it is worth remembering that compared to the number of species caught for food or destined for the curio and jewellery market, the number of animals captured for the aquarium is very small. And with the sophisticated equipment available today, many have a good chance of surviving in the aquarium at least as long as they would have in the wild.

Many countries have introduced legislation to control unsustainable exploitation. For example, the Philippines has banned the collection and export of all corals and giant clams. CITES (the Convention on International Trade in Endangered Species of Wild Fauna and Flora) regulates the levels of trade in wildlife, and takes appropriate measures if these become dangerously high. Hobbyists should ensure that their animals have been legally collected and exported from their country of origin.

Part Three presents a selection of invertebrates suitable for the aquarium. Warning panels advise against keeping certain species, either because they are dangerous to other species, or to the hobbyist, or because they do not survive long in captivity. Given a responsible attitude on the part of wholesalers, retailers and aquarists, there is no reason why the interests of conservationists and hobbyists cannot be reconciled. Finally, it is worth making the point that by observing their invertebrates, recording their findings and publishing their results, aquarists can contribute to the general fund of information about these fascinating creatures.

Left: *Neopetrolisthes ohshimai* the tiny anemone crab, safe among the tentacles of *Stoichactis gigas*. A well-stocked invertebrate aquarium can be an endless source of fascination and pleasure.

PHYLUM PORIFERA
SPONGES

Haliclona compressa
Red tree sponge

This bright orange-red species is very common in the Caribbean Sea and regularly imported. Most specimens are about 20cm(8in) tall, but larger examples are sometimes available and, with their interesting branched habit, they provide a dramatic splash of colour. It is important to select a specimen with the base attached to a piece of rock and with no white or pale patches on the arms. This species appreciates a reasonable water flow and, like all sponges, prefers somewhat dim lighting conditions. In too bright a situation, the branches often become covered with encrusting algae that choke the sponge and eventually kill it.

Adocia sp.
Blue tubular sponge

This intensely blue species is imported fairly regularly from Indonesia, but is never available in large quantities. Damaged specimens will turn progressively whiter as the living cells die, leaving behind the supporting structure. Once it has recovered from the initial transition from one tank to another, the blue tubular sponge is very hardy and grows surprisingly rapidly, particularly in an aquarium with fairly slow-moving water. Given suitable conditions it will also grow quickly from small pieces. It responds badly to large water changes and so, ideally, make these small and regular. Never remove any sponges from the water; if air pockets form within them they will decline and die.

In the wild, many small animals live within sponges and, on rare occasions, small crabs, shrimps and gobies are found in imported specimens. Other animals may live on the outer surface. Try to avoid introducing 'undesirable' subjects into the tank.

Above: *Adocia* sp.
Few other invertebrates are blue, so this animal provides a bold splash of colour in the aquarium. In slack or slow-moving water it will develop a branching form. An easy sponge species to maintain.

Left: *Haliclona* sp.
This striking Red Sea species is typically beige-pink but, as with many sponges, the colour is very variable. Avoid obviously damaged specimens and do not allow them to touch corals or anemones.

Right: *Axinellid* sp.
The Indo-Pacific orange cup sponge is readily available and easy to maintain if algae is prevented from smothering it. Remove any sediment that settles in the cup to prevent areas of decay developing.

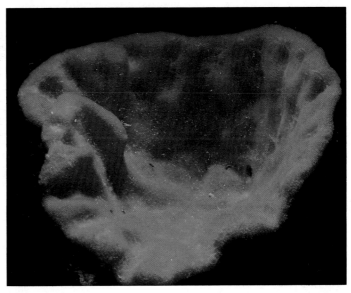

Axinellid sp.
Orange cup sponge

Sri Lanka, Singapore and Indonesia are the main sources of supply for this yellow-orange species with its distinctive cup or bowl shape. They often prove to be one of the hardiest species but, unfortunately, are often shipped in insufficient water, leaving the edges exposed to the air so that they may turn pale and begin to crumble. It is important to check that all sponges are intact before you buy them. Ensure that the tank is not brightly illuminated when you introduce this sponge and that there is sufficient water movement to prevent debris accumulating in the cup. Even with regular small doses of liquid feed, growth is slow. Several other types of yellow or orange sponges are regularly imported and most do well in the right conditions, but their shapes are less interesting.

PHYLUM COELENTERATA
SEA PENS, SOFT CORALS, SEA WHIPS, SEA ANEMONES, JELLYFISH AND HARD CORALS

Cavernularia obesa
Sea pen

Sea pens are an attractive and interesting group of animals, only rarely seen within the hobby. Their central tubular body is supported by a calcium 'spine' that resembles the quill of a feather, hence their common name. A distinct foot burrows into the substrate and anchors the animal in position in the fairly turbulent waters that they often inhabit.

During the day, the body, which may be orange, yellow, buff or white, is contracted. Throughout the night and occasionally during daylight hours, the body expands and previously hidden feathery polyps appear all over the animal as it starts to feed on plankton and organic detritus. Sea pens can give off an eerie greenish blue light when disturbed (see also page 18). Sea pens move very slowly through the sand and, in view of their habits, are not suitable for the total system aquarium with a very shallow substrate.

Right: *Dendronephthya rubeola*
The coloured cauliflower corals require a constant supply of fine food and good water movement to survive. Small specimens grow rapidly under good conditions.

Below right:
Dendronephthya klunzingeri
The typical vivid red colouring of this species ensures its popularity among hobbyists, but it is not easy to maintain in the home aquarium.

Below: *Cavernularia* sp.
The delicately branched polyps of this nocturnal species expand in the water current to capture minute particles of food. During the day, polyps and body contract.

Dendronephthya rubeola
Red cauliflower coral

This delicately branched species is one of the most attractive of the soft corals and the easiest of several lookalike species to maintain. *D. rubeola* lives on sand and mud sediments in the Indo-Pacific, where it anchors itself into position with a number of thick, fleshy, rootlike growths from its base. This method of attachment makes it much easier to collect this species than *Dendronephthya klunzingeri*, for example, which anchors itself to rocks and is easily damaged during collection.

During the day, these species contract into a red-and-white ball. The loose calcareous spicules that support the flesh like a disjointed skeleton project through the skin and make an uncomfortable handful. At night, or under subdued lighting, the animal takes in water to feed. Very large specimens reach over 1m(39in) in height, but more typical aquarium specimens are 15-20cm(6-8in) high. All the cauliflower corals feed on the very smallest particles.

Anthelia glauca
Pulse coral

The pulse corals are a comparatively recent introduction, but as Indonesian imports become more common they are justifiably increasing in popularity. *Anthelia* and the closely related *Xenia elongata* are two of the easiest corals to keep and both can be expected to spread and multiply in the aquarium.

   Anthelia consists of a cluster of very feathery polyps that join at the base to form a foot anchoring the coral to a rocky substrate. *X. elongata* is similar but more treelike, the body splitting into branches that bear the polyps. The common name for both these corals comes from the continual rhythmic opening and closing of the polyps as they feed. Both species require adequate lighting and appreciate a good current of water.

Below: *Anthelia glauca*
This attractive coral is highly recommended for the invertebrate aquarium. It spreads quickly, colonizing neighbouring rocks.

Right: *Xenia elongata*
Although very similar to *Anthelia glauca*, this species of soft coral has a more branched appearance. Provide the same care.

Above: *Sarcophyton* sp.
Here, this striking leather coral is shown with its polyps retracted – a natural defence mechanism.

Below: *S. trocheliophorum*
This mushroom-shaped coral expands its polyps to feed. With the appropriate diet and in good conditions, it grows steadily.

Sarcophyton trocheliophorum
Leather coral; Elephant ear coral

The leather corals are very widespread throughout the Indo-Pacific tropical seas and *S. trocheliophorum* is the most frequently imported. The common name comes from both the texture and colour of the animal when it retracts its polyps.

In the wild, they can form soft, undulated plates over 1m(39in) in diameter, the top surface clothed with a carpet of 1cm(0.4in)-high, delicate polyps. One of the best forms for the aquarium grows as a convoluted mushroom. Given good lighting, they fare very well and can be expected to increase in size.

When disturbed, the polyps retract and it is not unusual for them to take several days to re-open when, for example, they are transferred from one aquarium to another. This species can be strongly recommended as a first coral for the beginner. The best specimens are attached to a small piece of rock. Make sure that the base is undamaged, with no decomposing white areas. Offer small regular doses of a liquid feed and use a small-bore siphon to remove any sediment that accumulates within the 'mushroom'.

Muricea muricata
Sea whip

This Caribbean species is one of many that produce a cluster of 'finger', or whiplike, extensions. Sea whips are found throughout the tropics, usually in areas of strong water movement, and appear in all the colours of the rainbow. Most are fairly easy to maintain, the thicker fingered species having proved generally hardier than those with very thin branches.

Like leather corals, the best specimens are attached to a small portion of stone or coral. This ensures that the base is not broken and allows you to position the animal in a water current without risk of the soft flesh rubbing against the rockwork. When buying specimens, check that all the fingers are intact and that there are no exposed areas of dark chitinous skeleton. Bacterial infection can easily begin at such sites.

In nocturnal sea whips, the mat of small polyps only emerges at night. Offering diurnal species a fine liquid feed is an easier matter. Under ideal conditions, sea whips may grow 2.5cm(1in) a month.

Gorgonia flabellum
Sea fan

Sea fans are very closely related to sea whips, but in this species one main branch grows out in a very flat plane, the myriad small offshoots linking together to form a lace-fan appearance. Sea fans are particularly common on Caribbean reefs and at one time many were collected for the curio trade. Fortunately, this practice is now greatly reduced, but dried specimens of sea fans and sea whips are still occasionally offered as aquarium decoration.

When properly cured (i.e. made safe for aquarium use), sea fans should look like black lace. Before cleaning, they are often yellow or pink and should never be used in this state, as the dried tissue will rot in the tank and pose a major pollution problem. Although sea fans are more difficult to maintain than sea whips, they can survive in captivity, so let us hope that the dried sea fan trade will soon be a thing of the past.

Small, 15cm(6in) specimens are best for the aquarium as larger animals are difficult to transport without damaging the tissue.

Above: *Plexaurella* sp.
An impressive sea whip in its natural Cuban habitat, with *Pseudopterogorgia* behind it.

Left: *Muricea muricata*
A readily available Caribbean species, shown with an encrusting *Xenia* sp. growing on finger coral.

Right: *Muricea* and *Gorgonia* sp.
Colourful sea whips and sea fans are home to a spotted cardinal (top) and a birdnose wrasse.

Below: *Gorgonia ventalina*
This close-up view clearly shows the structure of a typical sea fan. Provide strong water movement.

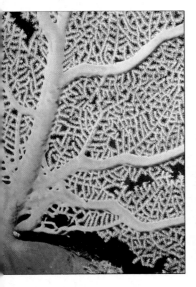

Pachycerianthus mana
Fireworks anemone; Tube anemone

The cerianthid anemones are an interesting group found in all the world's warmer waters. Their chief characteristic is that they live within a tube formed of mucus and detritus gleaned from the soft substrates they inhabit. They also have many very long, thin tentacles and a very powerful sting.

P. mana is an Indo-Pacific species that often has banded tentacles, while other close relatives show colours ranging from pure white, through yellow to maroon to near-black. Here again, the strength of their sting precludes them from acting as hosts for clownfish. In addition, their 20cm(8in)-long tentacles make them a threat to neighbouring invertebrates and fishes.

Cerianthids are night feeders, usually remaining in their tubes during daylight hours. They are easy to feed on finely chopped fish and shrimp every other day, but in view of the risk they pose to other animals, think carefully before introducing them to a well populated aquarium.

Left: *Pachycerianthus* sp.
The fireworks anemone lives up to its common name in appearance and can deliver a powerful sting.

Condylactis gigantea
Caribbean anemone

This long-tentacled anemone is the most popular and commonly exported species from the Caribbean sea. The body can be white, brown or pink, and the tentacles are usually pink or white with a more intense pink tip.

Condylactis are very easy to keep, requiring only moderate lighting and a steady, but not vigorous, water flow. They are easily fed by dropping small pieces of fish or shrimp among the tentacles once or twice a week. As a general rule, do not feed anemones with a liquid invertebrate diet.

Despite their pleasant colouring and ease of maintenance, however, *Condylactis* are not as popular as the Pacific anemones, because the various *Amphiprion* clownfishes will only very rarely set up home within their tentacles.

Left: *Pachycerianthus mana*
The tentacles of the tube anemone are lined with poisonous stinging cells, while the mouth is ringed with shorter tentacles. Not surprisingly, clownfishes will not use this family of anemones.

Below: *Condylactis gigantea*
The common Caribbean anemone is both inexpensive and hardy. It is available in a variety of attractive colours and, though easy to feed, rarely attains a diameter of more than 15cm(6in).

Heteractis magnifica (formerly *Radianthus ritteri*)
Purple base anemone

Heteractis magnifica is one of the large anemones that regularly plays host to clownfishes. Previously known as *Radianthus ritteri*, it was one of the few whose scientific name seemed to cause no confusion! Unfortunately, the new name *Heteractis magnifica* (Dunn 1981) has been very slow to achieve the same familiarity.

This anemone is found throughout the Indian Ocean and the Indo-Pacific region. Specimens from Sri Lanka typically have a purple body with buff or light brown tentacles. A more attractive form with a scarlet body and pure white tentacles is exported from Kenya. Both are easy to keep, but in good conditions they may reach 70cm(27.5in) in diameter and quickly grow too large for all but the biggest aquarium.

At first sight, *H. magnifica*, like all anemones, appears rooted in one position, but this particular species has an annoying habit of climbing slowly up the tank glass, thus presenting a rather unattractive view to the aquarist. By gently teasing the foot loose with the ball of the thumb, you can easily move them, but take great care to avoid tearing the very delicate flesh.

Heteractis aurora (formerly *Radianthus simplex*)
Sand anemone

The sand anemone's scientific name has also undergone a change (Dunn 1981). This greyish white species is easily distinguished by its tentacles, which are thickened to give a ringed appearance and have a tendency to lie flat against the surface disc. These are among the commonest anemones in their Indo-Pacific home range.

As their common name suggests, they are happiest placed on the substrate, rather than on rockwork, and will often anchor their foot through the sand and onto the undergravel filter plate or base glass of the tank. When disturbed, they can rapidly deflate and disappear beneath the substrate, thus escaping the attention of predators.

H. aurora is a small species, rarely more than 15cm(6in) in diameter, and is easy to feed with small pieces of fish or shrimp once or twice a week. Although easy to maintain for long periods, it is often thought to have only limited attraction for clownfishes. However, several *Amphiprion* species have been recorded with this anemone and *Amphiprion clarkii* have been observed with *H. aurora* in reef areas off Borneo. It is a very suitable species for the beginner or the hobbyist on a limited budget.

Above: *Heteractis magnifica*
In this purple-bodied example from Sri Lanka, most of the tentacles have been enfolded by the body. Popular with clownfishes.

Left: *Heteractis magnifica*
The rich red body is typical of specimens from Kenya. The starfish *Pentaceraster* just visible (top) poses no threat to the anemone.

Right: *Heteractis aurora*
This species is common on coral gravel beds within lagoons and at the edges of reefs. Here, the foot is buried, leaving only the ringed stinging tentacles exposed.

Aiptasia sp.
Rock anemone

This small, almost transparent, species is a common accidental introduction on good-quality living rock. Initially, the rock anemone is a welcome addition to the aquarium, but it reproduces very rapidly and can sometimes reach plague proportions. In this case, it can pose a real threat to small shrimps, gobies, blennies etc., and you must make an effort to control the population. Take heavily populated rocks from the aquarium, dip them in boiling water and scrub them clean before returning them to the tank. Inspect new decor carefully before adding it to the tank.

Left: *Aiptasia* sp. Remove from the tank as soon as possible.

Heteractis sp.
Red gelam anemone

In this species we have another example of the confusion that exists in the specific names for anemones. Although it is considered here as a *Heteractis* species, gelam anemones are structurally different to the malu anemones with which they share their family name. Gelam anemones have shorter tentacles – often with swollen tips – hence the alternative common name of bubble anemones.

Their small size – up to 20cm(8in) – and good colouring makes them justifiably popular. The most attractive gelam anemones have rusty red tentacles and the very best have purple bodies. They are easy to keep, feeding on small pieces of shrimp or shellfish, and most clownfishes will use them.

Gelam anemones have a tendency to roam around the aquarium, but can be encouraged to settle if they are placed in an opened clam shell or between two scallop shells wedged into the substrate. They usually confine themselves to the lower half of the tank, unlike *H. magnifica*, which often seeks out the highest point.

This species is particularly soft bodied; check that there are no tears in the body, as these usually prove fatal.

Above: *Heteractis* sp.
This large anemone is often a home for skunk clownfishes *Amphiprion perideraion* and *A. sanderocinus*. In this view, most of the tentacles have been enfolded after feeding.

Left: *Heteractis* sp.
A typical feature of the bubble, or gelam, anemones are the swollen tips to the tentacles. The body may be red-purple or brown and is normally hidden among the rocks.

Right: *Heteractis malu*
A pair of chocolate clownfishes *Amphiprion xanthurus* drive an encroaching common clownfish, *Amphiprion ocellaris* away from their home anemone. In the wild, clownfishes rarely venture far from their anemone and even lay their eggs under this protective mantle.

Heteractis malu (formerly *Radianthus malu*)
Malu anemone

Until recently, this species was widely known as *Radianthus malu* and will be much more familiar to experienced aquarists under this name. Malu anemones are imported in large numbers from Singapore, Indonesia and the Philippines, and are one of the staples of the hobby. *H. malu* is very attractive to clownfishes and an ideal species for most invertebrate aquariums.

Specimens are available in sizes ranging from 10 to 40cm(4 to 16in) in diameter, but they are capable of growing even larger. Three colour forms are available: pure-white, creamy yellow and light brown, but all display the distinctive purple-red tips to the tentacles, which are regularly tapered, up to 5cm(2in) long and evenly spaced across the disc.

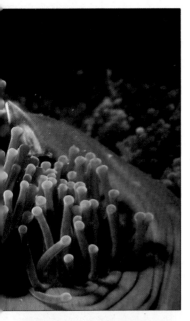

Given intense illumination, the lighter forms will turn brown with the development of the zooxanthellae algae that supply much of the anemone's nutritional requirements. In most situations, you can offer *H. malu* a similar diet to *H. aurora* and, like all anemones, it will benefit from regular additions of a vitamin supplement.

This species is more likely to stay where you put it than some others, but it is quite capable of moving if your choice of site is not suitable. When introducing this, and other anemones, to the tank, ensure that they do not rest 'face-down' on the sand. Anemones 'breathe' through the tentacles and quickly die if water movement around the tentacles is restricted.

Remember that all anemones accumulate waste products within their body cavities. To void these, they periodically collapse, pumping out the stale and polluted water within their body and often producing a stream of brown mucus at the same time. Occasionally, a sizeable anemone may shrink to the size of a golf ball. This is not a matter for concern, provided it does not happen more than once a day and the anemone does not stay closed for more than 24 hours. In this event, it may be taken as a sign that a major change of the aquarium water is overdue. Many anemones will contract at night, or when lacking in sufficient illumination.

Anthopsis koseirensis
Pink malu anemone

This attractive pinkish purple species is structurally very similar to the preceding species and is widely considered to be a colour form of *H. malu* – hence the common name. *A. koseirensis* is rather less common in the wild than *H. malu* and commands a higher price, but its appealing coloration, ease of maintenance and the willingness of all species of clownfishes to form a symbiotic relationship with it justifies its cost. It is particularly attractive if the illumination is supplemented with a red-enhancing light.

In the past, many anemones were artificially coloured by immersing them in a solution of food colouring to produce blue, green, orange and scarlet specimens. While the food dye itself appeared to cause no obvious problems, few specimens survived long after this treatment – possibly because of the effect it had on the colour of light reaching the vital zooxanthellae. Fortunately, this practice seems to have died out, but you should treat with suspicion any malu anemones in colours other than purple-pink, white, pale yellow and pale brown.

Below: *Anthopsis koseirensis*
This beautiful pink species is popular with hobbyists and clownfishes alike. Here, a pair of common clownfishes *Amphiprion ocellaris* are in residence, but other related species find this anemone equally attractive.

Stoichactis gigas
Carpet anemone; Blanket anemone

As the scientific name would suggest, this is one of the largest species of sea anemone, with wild specimens approaching 1m(39in) in diameter. Smaller specimens are popular with aquarists, but are not as readily available as in previous years, when large numbers were exported from Sri Lanka. Most were white or pale brown, but there was a steady supply of blue, purple and fluorescent green specimens. These are now rare in the hobby and command high prices. Nonetheless, this is a very hardy and long-lived species and worth seeking out.

All anemones have stinging cells with which they catch plankton, small fish and crustaceans, but this is one of the few in the hobby that can sting man. The short (1cm/0.4in), densely packed tentacles feel very sticky when touched; their effect is one of multitudes of barbed stinging cells being fired into the flesh. Stings on particularly sensitive areas, such as the wrists and inside of the forearms, can produce an annoying rash that may last several days or, on rare occasions, considerably longer.

In view of this stinging potential, be very careful when servicing the aquarium – and do not house *Stoichactis* anemones with other families of anemones nor allow them to come into contact with other invertebrates.

Feed this species in the same way as the gelam anemones, but you may find it necessary to wriggle the food into the tentacles to elicit a stinging response before the food is passed to the central mouth and engulfed.

Above and below:
Stoichactis gigas
This giant among the sea anemones has a powerful sting, so avoid introducing it to an aquarium with corals or other anemone species. Nevertheless, it is popular with clownfishes. Green and brown specimens are often available.

Rhodactis spp.
Mushroom polyps

Mushroom polyps – or mushroom anemones, as they are sometimes known – are available as colonies clustered on small pieces of rock. They have been one of the staple additions to the invertebrate tank for many years, but are still justifiably popular, being easily maintained even by the most inexperienced hobbyists.

There are many species; most of them measure 2-5cm(0.8-2in) in diameter, although a few giants can reach 15-20cm(6-8in), and these are very dramatic. Most species are green or brown, and often the two colours are combined in radiating stripes. Reddish brown and blue specimens are available from Indonesia but are somewhat expensive. It has been found that these red and blue pigments act as a light filter to prevent the animal being 'sunburnt'. If the aquarium is not brightly illuminated, these attractive colours disappear, allowing more light through to the zooxanthellae, but resulting in a drab, brownish animal.

A particularly attractive species, with a rather tufted appearance to its pale bluish green polyps, is often found on living rock imported from the Caribbean.

Below: *Rhodactis* sp.
There are many species and colour forms of *Rhodactis*, but all have the typical mushroom shape. Highly recommended for well-lit tanks.

Right: *Zoanthus sociatus*
The brown form of this species, growing on algae-encrusted coral rock. Bright yellow Indonesian species are sometimes available.

Zoanthus sociatus
Green polyps

There is a huge range of *Zoanthus* species and their relatives, all of which are generally sold under the title 'polyp colony'. The zoanthids mark something of a halfway house between the corals and the sea anemones. The polyps are usually found in clusters and are often interconnected at the base. However, they have many more tentacles than the true corals and no calcareous skeleton.

All the 'polyp colonies' are among the easiest coelentrates to keep. Most require reasonable water quality and good lighting, but very little else. In good conditions, mature polyps often multiply by budding new, small polyps at the base.

Most species are coloured in various shades of green, brown and beige, or combinations of the three, but several bright yellow types are regularly available.

Cassiopeia andromeda
Jellyfish

Jellyfish are familiar to beachcombers throughout the world, but virtually unknown within the hobby. *C. andromeda* is the only species ever offered for sale and the only one suitable for the aquarium. It is found throughout the Red Sea and the Indo-Pacific region and can reach up to 30cm(12in) in diameter. Only small specimens, up to 8cm(3.2in), are generally available.

Most jellyfish trail their arms behind them as they swim, but *Cassiopeia* spends much of its time lying on the substrate, with its tentacles upwards, wafting in the current. Among the tentacles are small bladderlike growths, filled with zooxanthellae that provide much of the animal's food. It therefore needs intense lighting, and you can supplement the diet with newly hatched brineshrimp and a good liquid feed. However, even with the best care, it rarely lives long in an aquarium and should be left to experienced aquarists.

Below: *Amplexidiscus* sp. One of the giant forms of the mushroom polyp, often known as elephant ears. It may reach up to 20cm(8in) in diameter.

Below right: *Cassiopeia andromeda* This jellyfish is lying on the sand, pulsing water over its tentacles to extract particulate food.

Goniopora lobata

G. lobata is common throughout the Indo-Pacific and is one of the most frequently imported corals. Its attractive feathery appearance, combined with its usually low price, ensures that this is often the first stony coral that many aquarists keep. Unfortunately, it is not necessarily a good choice, since it demands perfect water conditions and very good lighting.

Without sufficient light, the zooxanthellae will not function properly, and with too much food in the water, the polyps will not expand. Furthermore, *Goniopora* species are easily damaged. The polyps are not individual animals, rather outgrowths from the skin covering the ball-like skeleton. When the animal inflates with water to expand the polyps, the skin is stretched taught, becomes thin and is easily punctured. Sea urchins, sharp-footed crustaceans and tumbles from rocks are major causes of such punctures, which usually lead to a persistent infection that quickly engulfs the coral.

There are several species of *Goniopora*; one of the most attractive has rather thin, creamy polyps with purple centres. Treat them with great care. The *Porites* corals of the Caribbean and Pacific are closely related, but rarely imported and just as difficult to maintain.

Galaxea fascicularis
Star coral

This distinctive coral has a very attractive skeleton. Small stubs arise from the central core, with polyps sitting on each of the star-shaped ends. Singapore is the main source of this brown species, but as there are often considerable losses in transport it has not achieved any great popularity.

The star coral is by no means impossible to keep, but it requires similar conditions to *Goniopora* and cannot be recommended for the newcomer to the hobby.

Above: *Goniopora lobata*
Stony corals should be erect and well expanded. If the flesh is damaged, greyish brown areas of bacterial infection may appear.

Left: *Galaxea fascicularis*
Star coral is rarely seen in good condition, but here the colour is good and the polyps extended.

Right: *Platygyra* sp.
A close-up view of a Red Sea species of brain coral. The green colouring results from symbiotic algae living in the body tissues.

Far right: *Diploria strigosa*
A very large Caribbean species of brain coral in its home waters off St. Lucia. Small specimens make excellent aquarium inhabitants.

Leptoria spp.
Brain coral

Brain corals can reach massive proportions, but small specimens are strongly recommended for an aquarium with good lighting. The optimum size for aquarium specimens is about 10-15cm(4-6in); larger ones tend to be damaged in transport. Most species are various shades of brown, but some are a vivid green and others a pale purple-pink. Brain corals receive much of their food from the action of symbiotic algae, but they will accept small pieces of fish or shellfish meat lightly dropped onto them.

Occasionally, you may see small tentacles around the edges of the sinuous channels of the brain coral. These can be extended to sting nearby corals and are particularly prominent in the closely related Caribbean species *Meandrina meandrites*.

Favites sp.
Moon coral

These attractive dark green corals are closely related to the brain corals, but here the meanders are subdivided into numerous single pockets, each housing a polyp. The polyps are interlinked, which causes a major problem with this type of coral. All too often the specimens received by importers are merely pieces hacked from large coral heads and in these cases, even with the best care, they almost invariably die. *Favites* are among the most light sensitive corals, but can take supplementary food in the form of finely chopped shrimp. If decorative algae are growing in the aquarium, make sure that leaf fronds do not rob the coral of light. Moon corals fluoresce in ultraviolet light.

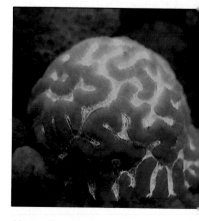

Acropora palmata
Elkhorn coral

The various species of *Acropora* are by far the most important corals, as they make up much of the coral reef structure itself. These huge and very heavy corals are almost never available to the hobbyist, since broken-off fragments soon die in the aquarium and small specimens are only rarely available. *Acropora* are among the most demanding of corals and should not be attempted by any but the most experienced hobbyist.

Much damage has been done to these corals by the predatory crown-of-thorns starfish and by man. Dynamite fishing, dredging for road and building stone and runway surfacing have decimated reefs. Aquarists should not slow down the natural regeneration of these reefs by seeking small specimens for the aquarium. Naturally killed specimens of this and related species, *A. cervicornis* (staghorn coral), *A. variabilis* (bush coral) and *A. pulchra* (plate coral) are acceptable as tank decor.

Above: *Favites* sp.
In this striking picture, a Red Sea species demonstrates the surprising ability of many corals to fluoresce under ultraviolet light.

Right: *Tubastrea aurea*
When it expands – which it does mainly at night – *T. aurea* is a most dramatic coral, resembling a bunch of golden chrysanthemums.

Below: *Favites* sp.
This more typical form of *Favites* is at home on the Great Barrier Reef. Undamaged, whole specimens do well in the invertebrate aquarium.

Below: *Acropora palmata* An impressive sight on a Caribbean reef.

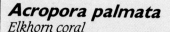

Tubastrea aurea
Sun coral

This vivid orange coral lives at the mouth of, or inside, caves and crevices in Indo-Pacific reefs. It is an extremely common species and large numbers are exported annually. Living as they do in a shady habitat, they have no need of symbiotic algae and rely on trapping food particles with their abundant yellowish tentacles.

This is an easy species to maintain, using the following technique. Place the coral in a shady spot and consider each polyp as an individual small anemone. If it is reluctant to open, tempt the polyps to expand and 'flower' by squeezing a shrimp head into the water. A few minutes later the polyps will open and you can feed each one a small piece of shrimp.

Newly introduced sun corals may not open for a week or more. From then on, if properly fed, you can expect the colony to expand by producing new polyps at the base of the mature ones. In the wild, colonies grow up to 50cm(20in) across, but 10cm(4in) is a more normal size in the aquarium.

There are several similar species, of which the Indonesian *Balanophyllia gemmifera* is a good choice for the aquarium. It is larger and even easier to feed than *T. aurea*. The related species *Dendrophyllia gracilis* has an attractive branchlike form and is only infrequently offered for sale.

Euphyllia sp.
Vase coral; Frogspawn coral

Considerable confusion still surrounds the nomenclature of many corals and, until recently, much classification work was based solely on the dead coral skeletons, with no consideration of the living animal. Much of this confusion is found in the *Euphyllia* group, which includes several well-known and popular corals.

The tentacles of all these specimens are very pronounced and comparable with those of sea anemones, although they vary in shape. Vase coral may have round or crescent-tipped tentacles, while those of frogspawn coral are semi-transparent and irregularly bubblelike. An exception is fox coral, in which the polyps are like a row of flat toadstools.

Euphyllia picteti
Tooth coral

Euphyllia picteti has only become widely available in the past few years, but is already firmly established as a favourite, both with newcomers to the hobby and the 'old hands'. Most specimens come from Indonesia and are quickly snapped up. When deflated they are fairly unprepossessing, but within a few hours of introduction to the tank they swell to four times the area of their supporting skeleton. The flesh is usually a fluorescent green, or occasionally blue, lined with anemonelike tentacles that frequently have vivid orange tips.

The green colouring is due to the coral's symbiotic algae, so strong lighting is essential. Once again, however, this coral will take sizeable pieces of food and, indeed, the tentacles pack a powerful sting. They are quite capable of raising a painful rash on a careless aquarist's arm, or killing any sessile invertebrates placed too close to them in the aquarium.

Far left: *Euphyllia divisa*
There are several species of
Euphyllia; this one has white,
knobbly tentacles and is known as
frogspawn coral. Provide good
light and offer brineshrimp nauplii.

Left: *Euphyllia fimbriata*
The best specimens of vase, or dog
tooth, coral are bright green, but
most of the available specimens
are brown or beige in colour.

Below right: *Plerogyra sinuosa*
This distinctive species can be fed
like a sea anemone. Here, a healthy
growth of *Caulerpa prolifera* algae
promotes good water quality.

Below: *Euphyllia picteti*
This beautiful green species may
have pink or orange tips to the
stinging tentacles. Popular, but
often dearer than related species.

Plerogyra sinuosa
Bubble coral

This very hardy and accommodating species can be highly
recommended as a beginner's stony coral. The common name is a
very apt description, for during the day the polyp mouths and
tentacles are covered in a mass of bubbles. These may be fawn
coloured, but are a pure snow white in the best specimens. At night,
the bubbles deflate somewhat and the coral erects flowing,
6cm(2.4in)-long stinging tentacles to capture small shrimps in the
wild. From the aquarist's point of view, the beauty of this coral is its
willingness to accept whole shrimps and pieces of fish gently tucked
among the bubbles. If regularly fed, a bubble coral can increase its
expanded diameter by some 50 percent within a few weeks.
Weekly feeding is sufficient and even this can be suspended if the
coral is given sufficient light.

The best specimens come from Sri Lanka and Indonesia, but
somewhat similar species, often greenish or light brown, are found
throughout the Indo-Pacific. Do not place any coral species too
close to one another.

Tubipora musica
Organ pipe coral

This species is much more familiar to aquarists as a dead skeleton for tank decoration than as a living animal. It is one of the few corals with a naturally red skeleton, a very attractive feature when not overgrown with algae. When it is alive, the top of the interlinked red pipelike skeleton houses a mass of short but active brown polyps. These pulse with the flow of water in similar fashion to *Anthelia*. The polyps are all interconnected and, as specimens are usually fragments of much larger pieces, their life in the aquarium is limited.

Above: *Herpolitha limax*
The strong light and moderate water movement in the shallows of the Great Barrier Reef suit this slipper, or hedgehog, coral well.

Left: *Tubipora musica*
The tubular structure of its red skeleton has given rise to the common name of this attractive coral. Aquarium specimens are usually broken from larger heads.

Right: *Heliofungia actiniformis*
The tentacles of this Australian plate coral are semi-extended, revealing the green-tinged body. An attractive Indonesian form has bright pink tips to the tentacles.

Millepora spp.
Stinging coral

This species is almost unknown in the aquarium, but is very familiar to divers who make a point of keeping well clear of its innocuous-looking flattened branches. *Millepora* are armed with a vast number of hair-triggered stinging cells, the poison from which can cause intense burning pain. In the wild, many small fish and shrimps find protection from predators among the branches of *Millepora*.

In the past, large quantities were collected, cleaned and dried for decorative purposes, as this coral has a naturally blue skeleton, which was considered highly desirable.

Right: *Millepora alcicornis* photographed in Florida.

Heliofungia actiniformis
Plate coral

The various species of *Heliofungia* are easily confused with sea anemones, since their circular or oval bodies are covered with long tentacles that completely hide the ridged skeleton. Most corals are a collection of polyps, but *Heliofungia* is a solitary polyp with one central mouth. Zooxanthellae can tint them green or pink. As well as deriving nourishment from the zooxanthellae, *Heliofungia* will take chopped fish and shrimp in small quantities.

Heliofungia fare best when placed directly onto a coral sand substrate where they can receive good light and a moderate flow of water. Do not position them on rocks, otherwise the delicate tissue around the edge of the coral may tear and open up a path for infection. When buying specimens, check that all the tentacles are erect and that there are no bald patches.

H. actiniformis is roughly circular, as are most of the related species, but *Herpolitha limax*, which sports many short, brown tentacles, forms a long oval.

PHYLUM PLATYHELMINTHES
FLATWORMS

Pseudoceros splendidus
Red-rim flatworm

This vivid red, black and white species is one of the few occasionally offered for sale. Although somewhat nocturnal in its habits, it usually does well if not subject to predation. The bright colours of this group of animals are believed to serve a protective function, and few fish will eat them because of their foul-tasting mucus. However, many crabs and shrimps will quickly devour, or badly damage, flatworms. There are many small and insignificant species; one of these – reddish brown and 3-4mm (0.12-0.16in) long – can reach plague proportions. Always remove it from a living reef aquarium, as there is no predator that will eliminate it naturally without damaging other desirable invertebrates.

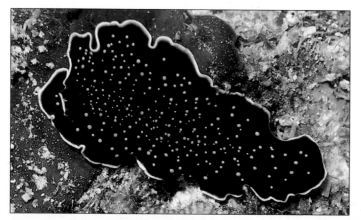

Left: *Thysanozoon flavomaculatum*
In the past, this attractive black, yellow and white species from the Red Sea was occasionally imported by accident among shipments of living rock. Now that exports from the Middle East are much reduced, it is rarely offered for sale.

Below: *Pseudoceros splendidus*
A most attractive member of a wide-ranging family of rarely seen animals. Most species are nocturnal, but *Pseudoceros splendidus* appears during the day, when its coloration, warning of a foul taste, deters predators.

Above: *Sabellastarte magnifica*
This cluster of featherduster worms illustrates the wide variety of colours available. Ideal subjects for newcomers to the hobby.

Below: *Sabellastarte magnifica*
This close-up shows the cilia, or 'feathers', used to trap fine particles of food, which are then channelled to the central mouth.

PHYLUM ANNELIDA
SEGMENTED WORMS

SABELLID SPECIES

Sabellastarte magnifica; S. sanctijosephi
Fanworm; Featherduster worm; Tubeworm

Among the huge numbers of featherduster worms and tubeworms imported each year from Singapore, Sri Lanka and Indonesia, these particular species are justifiably popular. The body of the worm is encased in a parchment tube buried in the substrate, with the feathery head extended for feeding. At the approach of danger, the feathery tentacles are very rapidly withdrawn into the tube. They are not fussy about their lighting requirements, can cope with comparatively poor water conditions and are easily satisfied with a simple filter-feeder food mix.

It is by no means unusual for these species to reproduce in the home aquarium. When this is about to happen, the first signs are usually evident early in the morning, when the animals emit smoky plumes of either eggs or sperm. The adult worms then often shed their feathery heads, as it makes no sense for them to eat the larvae that develop very shortly after the fertilization process is complete.

Tubeworms also shed their 'feathers' if attacked by predators. If this happens, remove the head but leave the tube in position, and after two or three weeks a short, stubby feathered head will normally reappear and eventually grow back to its former glory.

SERPULID COLONIES

Serpulid worms differ from sabellids in that they produce a stony tube and are usually considerably smaller. The most commonly available species lives in colonies, with its tubes embedded in *Porites* coral. The 1cm (0.4in) diameter worms are often very brightly coloured, with heads of red, blue, black, white and yellow – in contrast to the beige through brown to maroon heads of *Sabellastarte*. *Spirobranchus giganteus* the Christmas tree worm, is common in the Caribbean and the Indo-Pacific. It is so-named on account of the two branches of spiralling tentacles that emerge from the tube. Featherduster clusters from the Caribbean are small intertwined clumps of rocky-tubed serpulid worms, usually with red- or rust- coloured heads. They are easy to keep, but take care that these species do not become overgrown with algae.

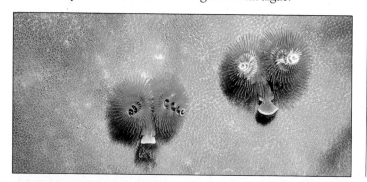

Above: *Serpulid* sp.
This large *Serpulid* species sports crowns that may reach 4cm(1.6in) in diameter. However, it is a fairly demanding variety and easily damaged in the aquarium.

Above right: *Serpulid* sp.
This small but attractive Caribbean species lives in small aggregations, the stony tube of one worm being cemented to that of its neighbour.

Above left:
Spirobranchus giganteus
This species appears in a startling range of colours; only pink and purple are rarely seen. Here, they are embedded in *Porites* coral. When disturbed, the worms retreat into their tubes and can close the entrance of the tube with the opercula visible in the picture.

Left: *Spirobranchus giganteus*
Aptly named the Christmas tree worm, the two spiral crowns are clear to see on this West Indian specimen. The crowns trap food and act as supplementary gills, withdrawing when threatened.

Hermodice carunculata
Bristleworms

Bristleworms are accidentally introduced into almost every long-standing invertebrate aquarium. They look like pink woolly-bear caterpillars, but are carnivorous scavengers that can damage or kill featherduster worms and some molluscs. They do not normally reach plague proportions unless there has been persistent over-feeding of the tank, but remove them at every opportunity. Do not pick them up with bare hands, as the fluffy-looking tufts along the sides are, in fact, extremely painful, needle-sharp calcium spicules. *Hermodice carunculata* can reach 25cm(10in) long, but there are many other smaller species that are just as undesirable.

Below: *Hermodice carunculata* Handle these pests with care.

PHYLUM CRUSTACEA
CRABS, LOBSTERS, SHRIMPS AND BARNACLES

Calappa flammea
Shame-faced crab

The shame-faced crab is one of the few typical crab-shaped crustaceans that you might consider for the home aquarium. A number of similar species are characterized by their over-developed but weak claws. These are normally held in front of the mouthparts with just the eyes peeping over – hence their common name.

Calappa are very efficient scavengers and will also break open and eat various types of molluscs. They are well camouflaged with algae growth on the carapace, or they may spend much of their time buried beneath the substrate – a habit that makes them interesting, rather than decorative aquarium inhabitants.

Macropipus sp.
Swimming crab

Small specimens of *Macropipus* sp. are occasionally imported from the Caribbean and when they measure 2–3cm (0.8–1.2in) across the carapace, they can be quite attractive. However, they are voracious predators, grow very quickly and soon become a threat to anything else in the tank.

In swimming crabs, the lower segments of the rear legs are flattened into paddles that propel them through the water at sufficient speed to catch small fish.

Below: *Macropipus* sp. A fast-growing predator.

Above: *Calappa flammea*
The shame-faced crab is not the retiring creature you might expect. It is a powerful omnivore, capable of devouring many species of sessile invertebrate in its quest for suitable food.

Right: *Stenorhynchus seticornis*
Within this limited family of spiderlike crabs, the Caribbean species shown here is the most readily available. Do not worry if a limb is lost during transport; it will quickly regrow.

Stenorhynchus seticornis
Arrow crab

The arrow crab gets its common name from the distinctly triangular, arrowhead-shaped body. This feature, together with its very long thin legs, results in a creature too closely reminiscent of a spider to appeal to many tastes. This is regrettable, as arrow crabs are attractive, very easily maintained aquarium subjects.

S. seticornis has a leg span of about 15cm (6in) and is generally well behaved, although it may pull at featherduster worms, since small, burrowing worms form a major part of its natural diet. In captivity, it will eat any meaty food and has the added advantage of being one of the few creatures that will happily consume the carnivorous, scavenging bristleworms (*Hermodice carunculata*) that are sometimes accidentally introduced into the aquarium.

A similar species, *S. lanceolatus*, with somewhat broader stripes on the body, is found in the Eastern Atlantic, while *S. debilis* is the only known Pacific form. Neither of these two species is currently available within the hobby.

Unless you have a very large aquarium, it is not a good idea to keep two arrow crabs together, as they almost invariably fight – the loser having all his legs removed before being eaten.

Neopetrolisthes ohshimai
Anemone crab

N. ohshimai is one of a small group of porcelain crabs that have evolved an immunity to anemone stings and, like *Amphiprion* clownfishes, can live among the tentacles of their venomous hosts for protection from predators. Measuring barely 2.5cm(1in) across the carapace, they are among the smallest crabs and make ideal aquarium specimens.

The crabs live in the same types of anemones as clownfishes and will use their well-developed claws on any clownfish that tries to evict them from their chosen home. For feeding purposes, however, the crab uses feathery projections on its jaw processes to trap particulate matter. As they are so small, anemone crabs are particularly vulnerable when changing their shells, so provide plenty of hiding places and do not house them with larger, more aggressive crustaceans.

The Indo-Pacific *N. ohshimai* has an irregularly spotted pattern that can vary depending on where it was found. *Neopetrolisthes maculatus* is densely covered with small chocolate spots on a white background to give it an overall pinkish appearance. The very rare *N. alobatus* from East Africa has widely spaced, almost circular, dark brown spots of varying sizes, producing a polka-dot effect.

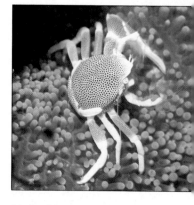

Above: *Neopetrolisthes maculatus*
An Indian Ocean species living on a *Stoicachtis* anemone. *N. maculatus* is regularly available and very popular with invertebrate keepers.

Below: *Neopetrolisthes ohshimai*
A particularly well-coloured pair of anemone crabs in a *Heteractis* anemone. They will feel more secure housed in fairly small tanks.

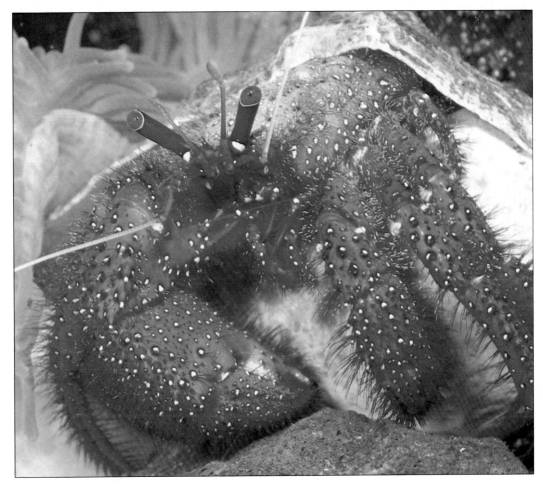

Above: *Dardanus megistos*
The red hermit crab is attractive
and interesting to observe but,
given its size and large appetite, is
better suited to a species tank.

Below: *Dardanus megistos*
A golden-yellow hermit crab
makes light work of carrying the
heavy shell it has made its home.

Dardanus megistos
Red hermit crab

Hermit crabs vary quite considerably in size and colour, from the
blue-legged hermits from Singapore and the tiny thumbnail-sized
species commonly shipped from the Caribbean to the giant *Aniculus
maximus*, which has attractive golden yellow legs but is a fearsome
predator that will devour anything that comes within reach of its
powerful claws. *D. megistos* is one of the largest species, with fist-
sized specimens by no means uncommon.

Unlike most crabs, the hermit crab's abdomen extends out from
the body, with no hard protective shell on the rear. It protects itself
by taking over the shells of various univalve molluscs – often by
eating the previous and rightful owner. Despite the weight of some
of these shells, hermit crabs are very active climbers and their
inquisitive nature endears them to many hobbyists. However, they
have very catholic tastes and a large hermit, such as *D. megistos*, is
capable of causing considerable damage within a well-stocked
living-reef aquarium. They are useful scavengers, however,
particularly in tanks with sizeable fishes, but will rarely fit into the
average aquarium set-up.

Pagurus prideauxi
Anemone hermit crab

Several species of *Pagurus* hermit crabs have gone one step better in their search for protection by actively encouraging certain species of stinging sea anemones to colonize their shells and ward off predators. Like all crustaceans, they periodically shed their hard outer skeleton, so be sure to include a few spare shells among the tank's decorations to provide new homes for the growing crabs.

Pagurus not only find new and suitably sized shells into which they can rapidly slip their delicate abdomens, but they also tease their anemones off the old shell and deliberately replace them on the new one. The anemones appear to accept this willingly because as the crabs rip up their food, many small pieces drift away and into the anemones' tentacles. *Pagurus* are large, destructive crabs and will only suit the aquarist looking for a 'one-off' speciality animal.

Above: *Pagurus prideauxi*
Another large, destructive hermit crab. By encouraging anemones to colonize its shell, it gains added protection from predators.

Right: *Lybia tessellata*
One of the most attractive of all the crustaceans. Best observed in a small tank where it is less likely to be attacked by larger animals.

Uca sp.
Fiddler crab

Until a few years ago, fiddler crabs were imported in considerable numbers. Today, their unsuitability for the marine aquarium is clearly recognized, but they still appear on the market from time to time.

Fiddler crabs have a small 5cm(2in) carapace and the male is characterized by its one hugely enlarged claw. This is so overdeveloped that it is useless for feeding and, instead, is waved at other males to drive them from the owner's territory on the mangrove mudbanks. The crabs live by chewing the mud, extracting organic matter and rejecting the remainder. As the tide encroaches, the crabs disappear down tunnels that retain a bubble of air to sustain them until the tide retreats.

Fiddler crabs of various species are very common throughout the Indo-Pacific; unless you are prepared to design a tank specifically for them, that is where they should remain.

Below: *Uca pugilator* Fascinating, but not an aquarium subject.

Lybia tessellata
Boxing crab

The small boxing crabs rarely grow more than 3cm(1.2in) long and make ideal aquarium occupants, particularly for small tanks, where they will not get 'lost in the crowd'.

Boxing crabs are the only examples of invertebrates known to use tools. While the anemone hermit crab, *Pagurus prideauxi*, merely shelters beneath anemones, these small crabs collect a tiny anemone in each claw and actively wave them at encroaching predators as a warning. Furthermore, even though these crabs use their first pair of walking legs to search the substrate detritus for food, they will happily collect food from the anemones. Only when they change their exoskeleton will *Lybia* deliberately release the anemones and carefully set them aside until the new shell hardens. Then they pick them up and press them into service once more.

There are several species of *Lybia*, but the nomenclature is in some confusion. All are attractive and very interesting animals, well worth a place in the tank.

Panulirus versicolor
Purple spiny lobster

This attractive purple and white banded species is the most attractive member of a large and commercially valuable family of animals and a worthy addition to the aquarium. There are major fisheries of its relatives in both the Caribbean and Mediterranean.

Very young specimens of *P. versicolor* are commonly imported from Singapore and Indonesia. They have a body length of 5-7.5cm(2-3in), the long, rasplike tentacles adding a further 10-15cm(4-6in). An adult body length of 20cm(8in) is by no means unusual. These efficient scavengers thrive in captivity on a diet of frozen fish and shrimp. Although not deliberately destructive, they can cause damage with their sharp feet, or by jerking backwards to evade a threat, either real or imagined.

Several other species of *Panulirus* are occasionally available, usually in shades of reddish brown, but these grow even larger and are suitable only for a very large aquarium. Like the purple species, their long antennae are easily broken in confined spaces and, although they will regrow with successive shell changes, the animal loses much of its appeal if these appendages are broken.

The slipper lobster, *Scyllarides nodifer*, is a close relative, but instead of antennae this species has well developed plates around the head to dig through the substrate seeking food. It is dully coloured and generally of more interest to the specialist.

Below: *Panulirus versicolor*
This strikingly marked species is just one in a large family of spiny lobsters, most of which are of commercial interest as food animals. The purple spiny lobster is very popular with aquarists, but make due allowance for its healthy appetite and rapid growth rate.

Above: *Enoplometopus debelius*
Shipments from Singapore and
Indonesia frequently include this
shy and retiring dwarf species.

Below: *E. occidentalis*
A naturally nocturnal species, but
E. occidentalis usually learns to
scavenge in daylight hours.

Enoplometopus occidentalis
Red dwarf lobster

This vivid red species is the most attractive of the various lobster
species and, with its relatively large claws, looks much more like
the typical fishmonger's lobster than members of the preceding
family. The red dwarf lobster grows to a length of about
12cm(4.7in) and looks very dramatic, but its largely nocturnal
habits ensure that you will generally catch only the occasional
fleeting glimpse of your specimen.

Enoplometopus is highly territorial and quickly despatches any
similar species and many of the more commonly available shrimps.
It is also quite capable of catching and killing small fishes,
particularly when these are 'dozing' at night. Think carefully before
introducing *Enoplometopus* into your aquarium, since removing a
particularly aggressive specimen may entail a complete strip down
of the tank at a later date.

The similar, but rarer, Pacific *E. holthuisi*, is slimmer and can be
easily distinguished by a white bullseye-like mark on either side of
the thorax. Two rather purplish pink species are imported from the
Indo-Pacific region and both are smaller than *E. occidentalis.*
E. debelius has a pale pink body, liberally covered with almost
round, purplish red spots. *E. daumi* has a pale purple-brown body,
becoming more richly purple towards the head and claws. These
two species are even more shy and retiring than their red relative.

Stenopus hispidus
Boxing shrimp; Coral-banded shrimp

The attractive boxing shrimp is one of the most commonly imported and justifiably popular species available to hobbyists. *Stenopus* are bold animals and, once settled into a new environment, will rarely hide for long. A large specimen with its 6cm(2.4in)-body looks very impressive as its long claws and delicate white antennae wave in the current. Unless you can be sure of buying a compatible pair, house only one specimen in the tank. However, *S. hispidus* generally mixes quite happily with other species of shrimp.

There are several species of *Stenopus* and as the hobby expands, so more appear on the market, albeit in very limited numbers. *Stenopus spinosus* is found in the Mediterranean and is essentially gold-yellow overall. The Caribbean is home to *S. scutellatus* which has red-and-white claws and tail, but a yellow thorax. It can be distinguished from the smaller *S. zanzibaricus* by the two scarlet dots near the mouth. *S. tenuirostris* from the Indo-Pacific is a small and lightly built species with a purplish blue thorax.

The giant of the family is the rare and very expensive *Stenopus pyrsonotus* from Hawaii, which grows at least 50 percent longer than *S. hispidus*. *S. pyrsonotus* is white throughout, except for a striking, broad, red band down the length of the body. The only similar species is the very small (2-3cm/0.8-1.2in) *S. earlei*, which has a pale body with a red stripe along each side, forming a 'V' shape at the tail. The base of the claws is reddish.

Stenopus are particularly vulnerable when they shed their exoskeletons and thus need plenty of hiding places to enable them to evade predators while the new shell hardens. Legs, claws or antennae are often lost or damaged, but these quickly regrow. All *Stenopus* are omnivorous, taking almost any commercial foods.

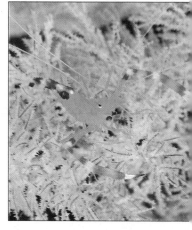

Above: *Stenopus scutellatus*
The yellow-bodied Caribbean boxing shrimp is rather delicate; do not keep it with larger and more aggressive *Stenopus* species.

Below: *Stenopus hispidus*
The bright colours of this popular boxing shrimp stand out in contrast against the green *Caulerpa* algae.

Above: *Lysmata amboinensis*
With its vivid red and white livery, it is easy to see why the cleaner shrimp is a favourite with aquarists.

Below: *Lysmata amboinensis*
A cleaner shrimp removing dead skin, parasites, etc., from a black velvet angelfish (*Chaetodontoplus melanosoma*) in the aquarium.

Lysmata amboinensis
Cleaner shrimp

The common name for these very attractive and sociable shrimps derives from their natural cleaning behaviour. In the wild, on Indo-Pacific reefs, they will pick parasites, damaged skin, etc., from many species of fishes, particularly moray eels and large groupers. The fish clearly appreciate these attentions and only rarely eat what would otherwise seem to be an attractive morsel.

Lysmata lack the dramatic claws of *Stenopus*, but many are attractively coloured. The common Indo-Pacific *L. amboinensis* has a scarlet back with a white stripe running from between the eyes to the base of the tail. The tail is marked with three white patches. The very similar but, in Europe, much rarer Caribbean *L. grabhami* has a white stripe running to the tip of the tail, which is edged, rather than blotched, in white.

Both species reach about 8cm(3.2in) in body length, and make very desirable aquarium inhabitants, being long-lived and easy to care for. They are particularly attractive when kept together in groups of four or five, when they may perform group cleaning activities on fishes.

Both species regularly breed in the aquarium, the females developing large quantities of green eggs under the abdomen. Unfortunately, the newly hatched shrimps usually provide a welcome addition to the menu of the tank's other inhabitants, or are quickly swept into the filter system.

Lysmata debelius
Blood shrimp

This intensely red and white spotted species caused a considerable stir within the hobby when it was first introduced in the early 1980s. Early specimens were all imported from Sri Lanka, but they have also been found in Indonesia and now, although still one of the more expensive shrimps, it is easily obtainable.

L. debelius is somewhat shyer than *L. amboinensis*, but after a few days adjusting to the aquarium conditions – and with the reassurance that there are plenty of convenient boltholes available – they soon settle down and do very well.

Several other *Lysmata* species occasionally appear for sale. Most are either banded or striped in combinations of white, pink, beige and red. *L. rathbunae* comes from the Caribbean, while *Lysmata californica* is found on the west coast of the USA. *L. vittata* is an almost transparent Pacific species, while the blotched pink and white *L. kukenthali* comes from the Pacific Ocean. A Mediterranean species, *L. seticaudata* and *L. wurdemanni* from the western Atlantic share similar red and white longitudinal markings.

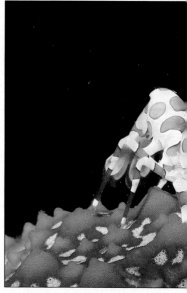

Hymenocera sp.
Harlequin shrimp; Orchid shrimp

For all practical purposes, *H. picta* from the Pacific and *H. elegans* from the Indian Ocean can be considered as one and the same species. Both are small – only 4cm(1.6in) long – with white bodies liberally blotched with red-brown or blue-brown respectively. Many of the body plates and claws are developed into broad, thin flaps.

In the wild, they generally live in pairs and are often available as such commercially. Although somewhat expensive, they would be very desirable aquarium residents but for one major drawback: their diet. In their natural environment, starfishes – and possibly the tube feet of sea urchins – provide their sole source of food. There have been occasional reports of them accepting other food in captivity, but the only certain route to success is a continual supply of small starfishes. When hungry, they seek out a starfish and lever it onto its back before breaking off and eating small sections. While one small starfish will suffice for a pair of shrimps for a week or more, this is rather too high a price, both morally and financially, for many aquarists to pay.

Above: *Hymenocera elegans*
A harlequin shrimp about to make a meal of a *Protoreaster* starfish. To keep this shrimp successfully, provide its specialized diet.

Right: *Rhynchocinetes uritai*
Candy shrimps are common in the wild, frequently imported and inexpensive. They make hardy and interesting aquarium subjects.

Left: *Lysmata debelius*
The aptly named blood shrimp is easy to maintain and long-lived in the aquarium. As supplies increase, prices should gradually reduce.

Below: *Lysmata seticaudata*
This attractive *Lysmata* species is only rarely available and, being rather shy, is unlikely to become as popular as others in the family.

Rhynchocinetes uritai
Candy shrimp; Dancing shrimp

This group of shrimps is distinguished by having a movable rostrum (head spine) and very protuberant eyes. There are many attractive species, but *R. uritai* is the only species regularly available. It is found throughout the Indian Ocean and Indo-Pacific regions and exported in large numbers, making them among the cheapest shrimps to buy. They do best in small groups, when they will lose much of the shyness they display if kept singly. Males have large but ineffective claws, so in view of their lack of defensive armament, do not house them with larger, potentially more aggressive crustaceans. On occasion, they will pester corals and anemones, but are generally harmless. Candy shrimps grow to about 3cm(1.2in) and will suit even a very small aquarium. They are ideal specimens for beginners to keep.

Saron rectirostris
Monkey shrimp

This interesting species and its relatives are comparatively recent introductions to the hobby, with specimens being shipped from Indonesia and Hawaii. They can be distinguished by the strong spines or hooks along the thorax and between the eyes.

S. *rectirostris* has purple legs and a pale body spotted with brown. *S. inermis* has greyish spots on the pale front portion, while the rear half of the body is more brown in colour. *S. marmoratus* is brownish green, with the body heavily coated with hairlike extensions. In all species, the males have a greatly elongated first pair of walking legs. In *S. marmoratus* these are attractively banded in reddish brown and buff.

They are good scavengers, taking any meaty food, but are essentially nocturnal and thus rarely seen during daylight hours.

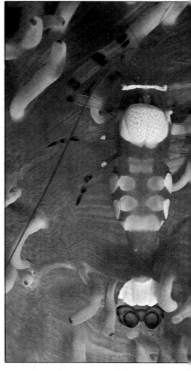

Above: *Periclimenes brevicarpalis*
Even with the protection of a stinging anemone, this little shrimp is often predated upon, particularly when shedding its old outer skin.

Left: *Saron marmoratus*
All the *Saron* species tend to be rather shy and here *S. marmoratus* is camouflaged among the fronds of a cactus algae, *Halimeda* sp.

Below: *Periclimenes holthuisi*
This glasslike anemone shrimp and its relative *P. pedersoni* are delicate species that do not appreciate boisterous neighbours.

Periclimenes brevicarpalis
Anemone shrimp

P. brevicarpalis is the commonest species in a family of very interesting and easily maintained shrimps. They have acquired their common name from their habit of living among the stinging tentacles of sea anemones for protection against predators, rather like clownfishes and anemone crabs.

These species are rarely more than 2.5cm(1in) long and are often almost totally transparent, which makes them very difficult to see if you are unaware of their presence. *P. brevicarpalis* is transparent with white blotches. It occurs throughout the Indo-Pacific region, where it often shares anemones with clownfishes.

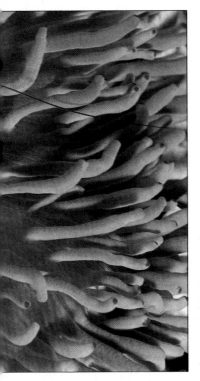

The very delicately formed *P. holthuisi* from the Indo-Pacific, and the Caribbean *P. pedersoni*, are glasslike with fine purple markings. *P. imperator* is found in the Red Sea and Indian Ocean and is very variable in colour. Bright red specimens have been found living among the gill tufts of large, similarly coloured sea slugs, such as *Hexabranchus*, the Spanish dancer. All will accept almost any small food items and will thrive in captivity. Do take care not to put them at risk from larger predatory species, however; most crabs and shrimps can stalk unharmed through anemone tentacles and an anemone shrimp would make a welcome addition to their diet.

Right: *Periclimenes imperator*
The colours of this species range from intense scarlet to almost all-white. Rarely available, but a welcome addition to the aquarium.

Above: *Odontodactylus* sp.
Powerful and deadly.

Odontodactylus spp.
Mantis shrimp

Unless you are prepared to devote an aquarium to this animal alone, avoid mantis shrimps. There are several families and species of mantis shrimps and all are extremely efficient predators, taking shrimps, crabs and fishes and damaging starfishes and featherduster worms.

Odontodactylus can grow to around 15cm(6in), and resembles a heavily armoured caterpillar. The head is densely spined and two clublike claws are ready to pound prey, quite literally, into pieces. A large specimen can split a 10cm(4in) diameter crab in half, with a force similar to that of a .22 bullet, strong enough to shatter the side of the aquarium.

Squilla spp. look more like the praying mantis insects after which the group is named, having fiercely barbed and hooked claws that they use to impale prey. Like pistol shrimps they are common accidental introductions; remove them as soon as you can but, as they are efficient tunnellers, this may not prove easy. Mantis are only rarely offered for sale and should be firmly ignored by the hobbyist.

Synalpheus sp.
Pistol shrimp

Small specimens of pistol shrimps are frequently introduced to the marine aquarium by accident, along with pieces of living rock. They often occur in the water canals of sponges. Most are pale brown or green, but there are a few very attractive orange and red species. However, they are all confirmed recluses and of limited interest to most hobbyists.

   Their common name comes from the pistol crack sound they are able to produce, a sound so similar to cracking glass that many a hobbyist has had a nasty shock! All pistol shrimps have one greatly enlarged claw, usually the right. By snapping this shut they produce a shock wave through the surrounding water, stunning the small shrimps that make up a major part of their natural diet. With a maximum body length of about 5cm(2in), they pose little, if any, threat to the other inhabitants of the tank and should not be confused with mantis shrimps.

Lepas anserifera
Gooseneck barnacles

Lepas species are occasionally available, but require large quantities of fine food, which can put a strain on the filtration systems of most marine aquariums. Small conical barnacles sometimes grow up spontaneously, or may be introduced on living corals.

Above: *Synalpheus* sp.
The greatly enlarged snapping claw of this pistol shrimp is clearly visible as it rests on a *Tridacna* clam. Normally a secretive species.

Below: *Lepas anserifera*
A small colony of gooseneck barnacles clustered on a piece of submerged driftwood. Common in the tropics, but rarely seen for sale.

PHYLUM CHELICERATA

Limulus polyphemus
Horseshoe crab; King crab

Limulus is the largest species, reaching 50cm (20in), and is found on mud and sand flats on the east coast of America. Small specimens are available on the American market, but in Europe one of the three western Pacific species is more likely to be on offer. These do not grow as large and make entertaining, if somewhat clumsy, aquarium specimens. They spend a large part of their time burrowing under the substrate and can play a valuable role in keeping the sand loose in undergravel filter systems.

Worms, algae and small shellfish form the natural diet of horseshoe crabs. They are unlikely to find sufficient food if left to scavenge in the aquarium, so provide small particles of squid, cockles and shrimps for these largely nocturnal animals.

Below: *Limulus polyphemus*
An adult horseshoe crab stranded upon the beach, where it has just laid its eggs. These interesting animals will prosper in captivity, given an adequate diet.

PHYLUM MOLLUSCA
SEA SNAILS, SEA SLUGS, CLAMS AND CEPHALOPODS

UNIVALVES

Cypraea tigris
Tiger cowrie

The tiger cowrie is one of a large group of sea snails of interest not only to aquarists, but also to shell collectors. Some cowries are extremely rare and command very high prices among collectors, but the tiger cowrie is not one of these. All the cowries have characteristic highly polished oval and domed shells, many of which sport very decorative patterns. This colouring is normally well concealed by the fleshy mantle – an extension of the body that folds up and over the shell. The mantle is usually decorated with tufts and tassles and inconspicuously coloured for camouflage.

Tiger cowries are commonly imported from Singapore at about 7.5cm(3in) long and are among the easiest invertebrates to keep. They tolerate less than perfect water quality, but demand good aeration. They feed avidly on *Caulerpa* species of seaweed and graze the less decorative hairy algae, but must have some animal matter; small pieces of fish and shellfish meat are suitable.

Below: *Cypraea tigris*
The spotted shell of the tiger cowrie is largely covered by the fleshy mantle, an extension of this sea snail's foot. *C. tigris* has a large but easily satisfied appetite.

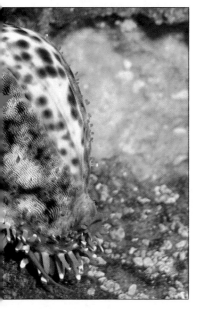

Left: *Cypraea arabica*
A map cowrie, its camouflaging mantle retracted, on the skeleton of an organpipe coral. A commonly available and attractive species.

Right: *Ovulum ovum*
One of the most striking univalve molluscs. Despite the difficulty of satisfying its specialized dietary needs, it is always in demand.

Below: *Cypraea* sp.
The heavily fringed mantle of this small species suggests that it is probably *C. nucleus*, an easily maintained and long-lived species.

Cowries have two major drawbacks; firstly, many are primarily nocturnal and hide under rocks during the day, and secondly, they are somewhat clumsy. They have a powerful foot that is not easily dislodged and often tumble corals and other sessile invertebrates from their allocated position in the aquarium. Additionally, some species and specimens develop a taste for both hard and soft corals.

Among the other regularly available species are *C. arabica* from the Indo-Pacific, a slightly smaller species with a netlike shell pattern; *C. pantherina*, which is somewhat similar to *C. tigris*, and *C. nucleus*, characterized by the furry appearance of its mantle.

Ovulum ovum
Egg cowrie

Despite its common name, this is not a true cowrie, but a member of the Ovulidae family. Superficially it is very similar to the Cypraeidae cowries, having a highly polished white shell and a very dramatic black mantle, liberally spotted with yellow. The egg cowrie is one of the most attractive univalves, but is more delicate than the cowries.

The main commercial sources for egg cowries are Sri Lanka and Indonesia, but as these animals are neither as common as cowries, nor do they travel as well, they are always considerably more expensive. They require optimum water conditions and you must take great care when acclimating this species to the tank. The main difficulty with their long-term culture is providing an adequate diet; most, if not all, specimens demand supplies of soft leather coral (*Sarcophyton* sp.) if they are to prosper.

Egg cowries can reach more than 10cm(4in) in length, but are usually sold at considerably smaller sizes. They are less nocturnal in habit than the true cowries.

Strombus gigas
Queen conch

The Queen conch is a very important commercial food animal in its native habitats and huge numbers of dead shells are exported for decoration and the curio market. It is one of the largest molluscs available to the hobbyist, reaching 25cm(10in) or more, and is thus likely to be of greater interest to the specialist. At one time, strictly enforced regulations put a limitation on the size at which *S. gigas* could be taken from the sea and no small specimens were available to the hobbyist. In recent years, however, there has been a huge increase in breeding and farming this species, and small specimens, up to 5cm(2in), are now obtainable on the American market and it cannot be long before they are more widely available.

This species is found on sand and sea-grass fields throughout the Caribbean, where it eats algae and detritus. Its attractive brown and pink shell should prove popular with aquarists, but make due allowance for the size of the full grown animal.

Cyphoma gibbosum
Flamingo tongue

This small Caribbean species grows about 3cm(1.2in) long and is the only other member of the Ovulidae to be regularly exported. It was once shipped in huge numbers, but now that studies have revealed more about its feeding habits, exporters have adopted a more responsible attitude. The flamingo tongue is found among the forests of sea whips and gorgonian corals that grow on many Caribbean reefs and depends on these species for food. Without them, a lifespan of four to eight weeks is the best that can be expected and it is therefore clear that this is one species best left in its natural habitat.

Flamingo tongues' shells are rounded at the tips and have a central ridge. *Volva* sp. from the South Pacific is similar, but the shell is sharply pointed at either end. Here again the species depends on a coral diet. The vivid colours of both species have no camouflage value and are probably protective.

Below: *Cyphoma gibbosum* Will not survive long in captivity.

Above: *Strombus gigas*
When disturbed, *S. gigas* retracts into its shell, as shown here. The heavy armour is proof against most would-be predators.

156

Below: *Lambis lambis*
The attractive underside of the shell. The top surface is normally camouflaged with algal growth – effective protection in the wild.

Conus spp.

The various members of the cone family are only rarely available, but are included here in view of their deadly potential. They are efficient fish predators and can indeed defend themselves against any predator, including man. They spend most of their time buried in sand with only the breathing syphon exposed. When a prey animal appears, the cone's proboscis shoots out and at the moment of contact one of the specially adapted radular teeth – in the form of a dart – is stabbed into the prey. The tooth is hollow and acts as a hypodermic syringe to inject a rapid-acting poison. If the target is a fish, the poison acts almost instantaneously; in humans the effects are similar to that of cobra venom. Fatalities are rare, but not unknown; victims lose control of muscle function and die of respiratory failure.

Species that prey on fishes, such as *Conus geographicus* and *C. striatus* are the most dangerous. *C. textile*, a particularly common species, is less likely to bite, but treat all the cones with very great respect. As these animals spend most of their time buried in the substrate, it would be all too easy for the aquarist to be bitten while performing one of the many routine aquarium maintenance tasks. It is grossly irresponsible for anyone to offer these animals for sale.

Below: *Conus purpurescens* A potentially deadly predator.

Lambis lambis
Spider shell

A number of Indo-Pacific species are sold under the name spider, or millipede, shell. All are characterized by the five or more long, thin and often sharp extensions to the shell, and all species have a very long and horny foot with which they are able to right themselves if they are inadvertently upturned.

Various other species and families of univalves appear on the market from time to time, but few are of interest to aquarists. Moon shells, olives, murex, tulips and whelks are all predators and have no place in a 'living-reef' set-up. Top shells, limpets and chitons are largely vegetarian and can be treated like cowries and conchs.

SEA SLUGS
Aplysia sp.
Sea hare

The sea hares are an entertaining, if not particularly attractive, group of sea slugs. Typically, they resemble a greenish brown lemon, with continually waving flaps along each side. They have earlike projections on the head, and it is these and their habit of grazing on seaweed that gives them their common name.

Most specimens offered for sale are shipped from the Caribbean, but very similar species are found throughout the world. They are by far the easiest type of sea slug to maintain for any length of time, although the first few days in a new aquarium can be a testing period. They require good water conditions and a continual supply of vegetable matter, preferably in the form of algae. If these few necessities are met, they can survive in captivity for two years or more. Since the invertebrate aquarium occasionally suffers from a plague of green algae, sea hares are one of the most useful and harmless answers to this problem.

The common Caribbean sea hare, *A. dactylomela*, can grow to over 30cm(12in), although smaller specimens of 6-8cm(2.4-3.2in) are most commonly seen. It is a strong swimmer, making good use of the parapodial flaps around its body.

Above: *Aplysia dactylomela*
One of the easiest nudibranchs to maintain, given large quantities of vegetable matter. They make short work of algae mats in the aquarium.

Below left:
Hexabranchus imperialis
A stunning example of a Spanish dancer crossing a sea fan. It may not survive long in the aquarium.

Below: *Hexabranchus sanguineus*
A Red Sea species, rarely found in the hobby. Best left in their natural habitat until more is known about their dietary requirements.

Hexabranchus imperialis
Spanish dancer

The Spanish dancer is one of the largest and most dramatically coloured nudibranchs available. It is imported in small but regular quantities from all parts of the Indo-Pacific, in sizes ranging from 6-15cm(2.4-6in). When at rest, or browsing over rocks, the mantle of the Spanish dancer is folded and marbled red, pink and white in colour. It is not until it swims that its full glory is revealed. The mantle unfolds to reveal an expanse of vivid crimson with a white border and then, with an action like a butterfly-stroke swimmer, it 'flies' through the water, more than earning its common name.

The similar, and even more dramatically scarlet, *Hexabranchus sanguineus* can be found on reefs in the Red Sea, but it is some years since this species was available commercially. Both species seem almost immune to fish attack and one small shrimp, *Periclimenes imperator*, takes advantage of this to hide among the gill tufts of *H. sanguineus*.

Spanish dancers will often lay eggs in captivity, producing a 3-4cm(1.2-1.6in) diameter rosette of pink gelatinous ribbon containing thousands of eggs. Unfortunately, neither these, nor the adult animal, generally succeed in the aquarium. The adults are reputed to be omnivorous scavengers, but from their limited survival rate it seems likely that a significant ingredient is missing from their diet. *H. sanguineus* is said to feed on sponges and sea squirts and has been observed feeding on the elephant ear coral, *Sarcophyton trocheliophorum*. These species can only be recommended for experienced hobbyists prepared to experiment to find a suitable diet for these spectacular creatures.

Chromodoris quadricolor
Striped nudibranch

The dorids are the largest group of nudibranchs and include some of the most vividly coloured animals found in the sea. *C. quadricolor* though striking in its black, white and orange livery, is by no means exceptional. Members of this group typically have two retractable tentacles on the head and a ring of gills, again retractable, towards the rear. Specimens rarely grow more than 6cm(2.4in) long.

Although commonly available and among the cheapest of sea slugs, their life expectancy in captivity is short. Again, the problem is to provide the correct diet. All are predatory, feeding on a wide range of sessile invertebrates, from sponges to barnacles, sea squirts to soft corals, and most seem limited to just one prey species. In the Red Sea, *C. quadricolor* feeds on the red sponge, *Latrunculia*. Until more is known of their requirements and a suitable alternative diet is available, they are best left in the wild.

Other available dorid species include *Gymnodoris ceylonica*, which lays strings of yellow eggs; *Polycera capensis*, a sea squirt feeder; the green, yellow and black *Tambja affinis*, and the sponge-feeding, white *Casella atromarginata*.

Below: *Chromodoris quadricolor*
The bright colours of these common nudibranchs probably serve to warn off predators. Note the retractable tentacles.

Right: *Phyllidia varicosa* (left) and *Phyllidia* sp.
These knobbed sea slugs do not have the exposed gill tufts typical of many of their relatives.

Above right: *Glossodoris* sp.
This attractive Indian Ocean
species is pictured here with an
orange cup sponge. A related
nudibranch, white in colour with
black spots, is known to eat blue
tubular sponge, so this specimen
may well be enjoying a meal.

Right: *Chromodoris lubocki*
The purple sea slug has recently
become a regular inclusion in
shipments from Indonesia and the
Philippines. This dramatic
6cm(2.4in) animal often lays eggs in
the aquarium, but there are no
reports of successful breeding.

Spurilla sp.
Spiny nudibranch

Spurilla is one of a large group of nudibranchs known as aeolids, most commonly found in cooler waters, although there are some tropical species. Most are fairly small, measuring up to 3cm(1.2in) and, although common as accidental introductions, are rarely available commercially. Small grey species similar to *Spurilla* are common on *Goniopora* coral, but generally go unnoticed. As accidentals they usually fare better than the dorids, as they are usually introduced on their food animals.

This group characteristically has two long tentacles on the head and numerous spikelike protuberances on the back. They are often brightly coloured, but without the striping common in dorids. They generally feed on coelenterates (anemones and corals), but some eat molluscs and fish eggs.

One particularly interesting species is *Glaucus atlanticus*. This blue-grey species is oceanic, living at the surface where it hunts various floating coelenterates, including the notorious man-of-war jellyfish. Not only is it immune to the jellyfish's stings, but it can store them within its own body to deter predators, which might eat the otherwise defenceless *Glaucus*.

There are many thousands of species of sea slugs and a considerable number make the occasional appearance on the market. Among the most common are the 'warty slugs' which are often covered in pimples, but show no external gills or tentacles. As a general rule, the duller green-brown species are longer lived than the more brightly coloured types.

Above: *Tridacna maxima*
Often imported from Singapore in sizes up to 30cm(12in) *Tridacna maxima* will make a striking centrepiece in any aquarium. The marbled pattern is due to the symbiotic algae within the flesh.

Above: *Tridacna crocea*
Vivid blue specimens such as this are justifiably popular with aquarists. Usually seen at about 10cm(4in), they can reach twice this length, but are slow growing, even in the best environment.

Left: *Pteraeolidia ianthina*
This attractive blue species is one of the many aeolid nudibranchs. The distinctive tufted gill processes are clearly visible as it glides past a yellow sponge, *Pleraplysilla* sp., on a Red Sea reef.

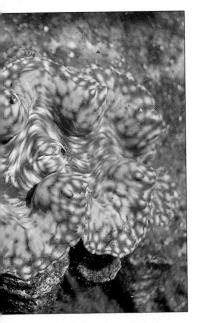

BIVALVES

Tridacna spp.
Giant clams

The various *Tridacna* species are probably the most popular family of bivalves, and justifiably so. The shell, particularly that of *T. crocea*, often shows deep flutes along the ridges, while the fleshy mantle of most species can be the most intense, almost fluorescent, blue and green. The largest species is *T. gigas*, the giant clam of Hollywood fame that can reach a weight of over 100kg(220lbs), and is particularly common on the Great Barrier Reef of Australia. Its shell is often covered with algae and coralline growths and the mantle is generally greenish brown. This species has been heavily over-fished and is largely protected.

Tridacna maxima and *T. crocea* are of more interest to the hobbyist and are regularly imported from Indonesia and Singapore. *T. maxima* can reach 30cm(12in) in length, and some of the most attractive species have green and brown striped mantles, while others may show a chocolate and cream blotched pattern. The most dramatically coloured are the blue specimens of *T. crocea*, which grow to around 15cm(6in).

All clams are filter feeders, drawing water through one siphon, filtering out planktonic organisms, and exhaling the cleaned water through the other. They are all heavily dependent on intense lighting. In fact, much of the colouring in the mantle is due to the zooxanthellae algae living in the tissue. These algae utilize sunlight for photosynthesis and produce the majority of the clam's food.

Below: *Tridacna gigas*
A giant clam in good condition. In poor light, the clam does not open fully, the zooxanthellae die and the colour fades. The clam should close when touched; if not, remove it from the tank before it decays.

Lima scabra
Flame scallop

This very attractive Caribbean species grows to about 6cm(2.4in) in diameter. The shell is unremarkable, but the body flesh is an intense scarlet. In the most popular of the two forms of *L. scabra*, the fringe of tentacles around the lip of the shell is also red, while in the other, the tentacles are off-white.

Position these scallops in a hollow or crevice towards the front of the tank where, hopefully, they will settle and attach themselves with wiry threads. Without an anchoring point, they may gravitate towards the back of the tank and be lost among the rocks.

Flame scallops are filter-feeders and have fairly heavy appetites. The commonest cause of death, other than predation, is long-term starvation due to insufficient supplementary feeding or an over-efficient filter system.

Flame scallops are a very popular food item for many animals, but are able to escape predators by clapping their shells together and using the resultant force to jet through the water. The Indo-Pacific lookalike *Promantellum vigens* has another useful defence system. The scallop's usual tactic is to flee the scene of battle, but if this fails, *P. vigens* defends itself with its tentacles. These are very sticky and easily detached from the body, and thus prove an irritating deterrent to many fish.

Occasionally, flame scallops reproduce in the aquarium and small clusters of spats – miniature, almost transparent 0.5cm(0.2in) diameter scallops – are found in caves and under rocks.

Below: *Spondylus aurantius*
An occasional import from Sri Lanka, but by no means common. The algae encrustation on the shell provides good camouflage from potential predators.

Left: *Lima scabra*
The vivid colouring of the flame scallop explains its popularity. This Caribbean species has lookalike relatives throughout the tropics.

Below right: *Spondylus americanus*
This species demands excellent water quality and an adequate supply of particulate food in the home aquarium. One of the more expensive marine invertebrates.

An interesting Philippine species has recently appeared on the market. It is similar in size and coloration to the red form of *Lima scabra* and has rippling luminous lines just inside the shell that flick on and off. We do not yet know what benefit these lines confer on the so-called flashing scallop.

Spondylus americanus
Thorny oyster

As its scientific name would suggest, this is a Caribbean species. Both valves of the shell have long, thornlike extensions and, when cleaned, the shell is very attractive and a favourite with collectors. When the animal is alive, the thorns are often heavily covered with growths of sponge, hydroids and algae. These camouflage the animal, which has prettily marked red and white flesh.

Spondylus aurantius is a very similar Indo-Pacific species with shorter thorns. This species is often coated with a vivid red sponge and, at 20cm(8in) in diameter, it is twice the size of *S. americanus*, and a very impressive animal.

Both species are found in caves and under overhangs and thus do not appreciate bright lighting. Unfortunately, they appear short-lived in the aquarium and, because of the pressure from shell collectors who pay very high prices for good specimens, they are generally too expensive to appeal to most hobbyists.

Very few other bivalves are deliberately added to the aquarium, but they are quite often introduced accidentally. Small specimens are often found growing on sea whips and sea fans, and many burrowing species are introduced with 'living-rock'. Some of the best Caribbean rock is heavily populated with burrowers and mussel species. Take great care to remove any air pockets from this type of rock, as tunnelling bivalves may otherwise die and cause very serious ammonia and nitrite pollution problems.

Above: *Lopha cristagalli*
The zig-zag cockscomb shell is often covered in red encrusting sponge, *Microciona* sp. A bivalve best left to experienced hobbyists.

CEPHALOPODS

Octopus cyaneus
Common tropical octopus

Octopi are the most advanced cephalopods and have lost all trace of their ancestral shells. Most species are less than 60cm(24in) in diameter – and none approach the horror-story dimensions prized by the early Hollywood film makers. *Octopus cyaneus* rarely reaches more than 30cm(12in) across and is typical of the small tropical species shipped in considerable numbers from the Far East. It is common on many reefs and easily captured by overturning stones under which it lurks at the water's edge at low tide. This and other species are able to control not only the colour of their skin, but also its texture. When at rest and relaxed, most octopi are fairly smooth skinned and their colour matches their background. When angered or frightened, they rapidly become much darker or lighter and their skin folds into eruptions resembling algae growth.

Like many octopus species, *O. cyaneus* is an ideal aquarium inhabitant for those prepared to make some effort on its behalf. Although not particularly light sensitive, all octopi demand perfect water conditions and will not tolerate any form of pollution. The aquarium must contain a number of suitable caves as hiding places and a tight-fitting lid. Octopi are notorious explorers, squeezing their boneless bodies through the smallest of openings. Many an octopus has met a dry and dusty end on the carpet, thanks to its owner's carelessness. Suitable tankmates include corals, sponges, featherduster worms, and some echinoderms. Crustaceans and fish, unless intended as food, have no place in the octopus tank.

Right: *Octopus cyaneus*
The powerful sucker-lined arms are clearly visible on this common octopus. It uses its tentacles to crawl, but is also capable of swimming. These highly intelligent creatures make fascinating aquarium subjects, but are best suited to a specialist tank.

Hapalochlaena maculosa
Blue ring octopus

This extremely attractive buff-brown species has a number of intense blue markings that usually form rings, hence its common name. It is very common on Australian and Indonesian reefs, where it occurs in very shallow water. Unfortunately, most exporters still list *H. maculosa*, apparently not realising the potential danger this animal presents.

All octopus capture their prey by enfolding it within their arms and then biting it with the parrotlike bill situated at the base of the tentacles. Most species have a toxic saliva that quickly immobilizes crabs, shrimps and fish. What the blue ring lacks in size, it more than makes up for with the toxicity of its venom. More than one careless beachcomber has been bitten and subsequently died after showing off his find. If you are aware of the risks, there should be little danger, but accidents do happen. Normal tank cleaning operations may put you at risk and if a blue ring were to climb from its tank, and a child picked it up, the consequences are easily imaginable. There is a case for making this animal subject to the Dangerous Animals Act and no home aquarist should keep one.

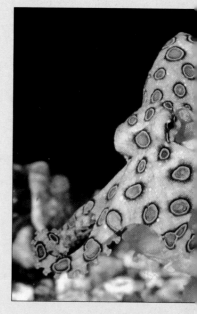

Above: *Hapalochlaena maculosa*

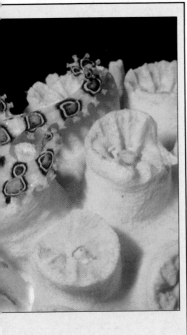

The octopus should be the last introduction to the tank, as it resents further disturbance. Allow it to become used to the dark, leaving the tank unlit for a day after the animal's release. Take the greatest care during this period; octopi are sensitive creatures and if badly upset will eject sepia ink into the water. This natural defence mechanism can cause major problems in the tank, and if the animal 'inks' in its shipping container it is likely to die.

A happy octopus is a greedy feeder, and laboratory tests have shown that it is quick to learn how to obtain food – even unscrewing bottle tops to get at the food within. Do not overfeed these ever-hungry creatures; they can generate more wastes than the average filter system is capable of dealing with in an acceptably short period. A regular feed of one or two shrimps or small frozen fish per day will comfortably satisfy all but the largest species.

It is not a good idea to house two octopi in any but very large aquariums, as they will usually fight. The female octopus is fertilized internally and occasionally a gravid specimen will produce fertile eggs in the aquarium. She may hang these from the roof of a cave or carry them about with her. The female does not normally feed during the incubation period and generally dies shortly after the eggs hatch, producing miniatures of their parents. The small hatchlings can be kept together and commercial producers in America are now supplying their country's growing number of public and educational aquariums.

Sepia plangon
Cuttlefish

The brittle supporting blade inside cuttlefish of the *Sepia* family is very familiar to birdkeepers. Unfortunately, cuttlefishbone is the closest most of us will ever come to having a cuttlefish, although *S. plangon* is a common species in the tropical Indo-Pacific, while *S. officinalis* occurs throughout the northeastern Atlantic and the Mediterranean Sea. The latter is a very common and popular animal in European public aquariums, where it can enjoy the large tanks that these very active animals require.

Unlike their relatives, the octopus, the squids are fast-swimming and very active hunters. Like an octopus, they have eight arms around the mouth, but they also have two longer and rapidly extendable arms to catch shrimps and small crabs.

Very little is known about the behaviour of the vast majority of squid species because they are difficult to catch undamaged and many live in very deep water. Those few species that have been studied display a variety of interesting behaviour patterns. Many are capable of very rapid colour changes and, in the case of *Sepia*, waves of colour wash over the body when they are excited or agitated. Many squid species show luminescent patches and it is believed that some species communicate with each other with a system of colour codes.

Very occasionally, small species, such as *Loliguncula brevis*, are imported from the Gulf of Mexico. This has very large pigment cells (chromatophores) that produce a kaleidoscope of reds and black. Given plenty of swimming room, it might prove viable in the home aquarium. All squid require optimum water conditions and are particularly sensitive to low oxygen levels. Most will not tolerate salinities lower than those of their native waters.

Right: *Nautilus macromphalus*
Polished nautilus shells are readily available, but the live animal is rarely for sale. However, greater numbers have recently been discovered off several deepwater reefs, so they may appear on the market with greater regularity.

Below: *Sepia plangon*
Cuttlefish are rarely imported alive and require a large aquarium and perfect water conditions. They are efficient hunters and cannot be trusted with fish or crustaceans. Small tropical species may show vivid blotched patterns, coloured red, yellow and black.

Nautilus macromphalus
Nautilus

For many years, Nautilus were thought to be extremely rare and to occur only in very deep waters. Recently, however, large numbers have been found and caught by research teams off Indo-Pacific reefs. It appears that they retreat to the depths, often hundreds of feet down during the day, but float up into shallower water at night and use their many tentacles to catch small fish and shrimps.

Very occasionally, the better wholesalers may obtain specimens of Nautilus, but the supply is so limited and the demand so great that their price is likely to remain beyond the reach of most hobbyists for the foreseeable future. Nonetheless, those specimens that have appeared have proved viable in the aquarium and, despite their essentially nocturnal habits, have many admirers.

PHYLUM ECHINODERMATA
CRINOIDS, BRITTLE STARFISHES, STARFISHES, SEA URCHINS AND SEA CUCUMBERS

Himerometra robustipinna
Red crinoid; Feather starfish

H. robustipinna, from Singapore and Indonesia, reaches about 18cm(7in) in diameter and is one of the most attractive species. *Lamprometra palmata* is one of several common brownish species that achieves a similar size, while others are occasionally available in shades from yellow through orange to black. All these species require very careful acclimation to the aquarium, as they react badly to rapid changes in salinity and pH. Do not house them with large, boisterous or 'pecky' fishes.

Crinoids feed mainly at night, climbing to a high point on the reef and then extending their arms to trap small particles of food falling from the surface of the sea. Nevertheless, in captivity they are also very decorative during daylight hours when they are unable to retreat to the cavities they would normally seek out in the wild.

Remember that these animals are very brittle (see page 44). If they are caught by a strong water current or attacked by other animals, one or more arms is easily broken. In the wild, these will regenerate very quickly, but in the aquarium they require perfect water conditions and a great deal of suitable food, in the form of

Left: *Himerometra robustipinna*
This vivid red species is shown here with its arms extended for feeding, as it nestles among the branches of a gorgonian soft coral.

pulverized shrimp or fish, algae fragments and newly hatched brineshrimp, if they are to recover.

In the wild, many small starfishes, shrimps and gobies live a well-camouflaged existence among the arms of crinoids; these 'extra' species are infrequent, but welcome, bonuses in the tank.

Left: *Comanthus bennetti*
The delicately branched arms of this yellow and black feather star trap planktonic food particles, as the current pushes water past.

Below: *Ophiothrix svensoni*
The central disc and five thin arms of this brittle starfish are typical of the group. Blue mushroom polyps are pictured in the background.

Ophiomastix venosa
Brittle starfish

Brittle stars are regularly imported from the tropics and are not expensive, but because they spend most of their time hidden from view they are not particularly popular. However, they are valuable scavengers, particularly in living-reef tanks, which often have many crevices that trap excess food. Furthermore, they will not harm sessile invertebrates, such as corals and featherduster worms.

Brittle stars have an efficient sense of smell to detect food and a surprising turn of speed (see page 44). If a promise of food awaits them, they will sprint from one end of the aquarium to the other in a matter of seconds.

Ophiomyxa flaccida is a typical smooth-armed Caribbean species that reaches approximately 15cm (6in) in diameter. This species is often seen entangled among the spines of *Diadema* sea urchins, where it is comparatively safe from predators.

Above: *Himerometra palmata*
A group of marbled feather stars gather on an Indonesian reef pinnacle as dusk falls. Most crinoids are more active at night.

Astrophyton muricatum
Basket star

Most imports of the fascinating basket star come from the Caribbean, but similar species are found in other oceans. They are very efficient feeders and require a good supply of food to thrive. They also need plenty of room to spread their 50cm(20in)-long arms, so they are clearly not animals for small tanks.

Given suitable conditions, these are long-lived creatures in the aquarium, but their strictly nocturnal habits limit their popularity to all but the most dedicated hobbyists. Their colours are generally restricted to greys, beige, browns and combinations of these.

Left: *Astroba* sp.
The delicate tracery of the arms of basket stars is well illustrated in this Red Sea species. At rest, the arms are curled into a tight ball.

Linckia laevigata
Blue starfish

The blue starfish is undoubtedly one of the most dramatically coloured of all marine invertebrates. While blue is a common colour among marine fishes, it is very unusual in invertebrates and the pure, intense, occasionally almost purple, blue of *L. laevigata* makes it look almost artificial.

This species has been regularly imported from Singapore, Sri Lanka and the Philippines almost since the beginning of the marine hobby, and is now often considered a little 'old hat' by experienced hobbyists. Nonetheless, the species has many factors in its favour. It is quite inexpensive, feeds well and, unlike some species, spends a considerable part of the day on view. But there is one particular caution; this species seems particularly prone to parasitization by a small species of bivalve mollusc that buries into the animal, usually from the underside of one of the arms. If left in place, it will ultimately penetrate the critically sensitive vascular system. These parasites are easily removed with a gentle thumbnail, but it is always wise to examine blue starfishes carefully before buying them and to reject any that show evidence of damaged skin.

Culcita novaeguinea
Bun starfish

The adult of this large 20cm(8in), Indo-Pacific species has a completely different body form from that of the typical starfish. As a youngster, it shows the typical five-pointed star shape familiar to us all, but as it matures the arms broaden and thicken to produce an almost regular pentagon. In addition, the body is comparatively thick and the final result is one of the heaviest starfish in the world.

At their smaller sizes, this and the related *C. schmideliana* make very attractive aquarium inhabitants but, unfortunately, they are not common and only rarely appear in the retailers' shops. The underside is generally plain, but the dorsal surface is often attractively marked with a pattern of dark raised tubercles.

Bun stars do best in lightly decorated tanks, as they will often wedge themselves among rocks and risk puncturing their skin, thus opening a route for bacterial infections. They are also somewhat clumsy animals and may dislodge and damage loosely positioned corals, etc. as they move around the aquarium.

Left: *Culcita novaeguinea*
Most bun stars have a beige-brown dorsal surface, but the undersides are often brightly coloured.

Right: *Culcita schmideliana*
Note the channels housing the tube feet (mostly semi-retracted) and the mouth in the centre.

Below: *Linckia laevigata*
Three bright blue starfishes cross a patch of the Great Barrier Reef in search of edible detritus. Most specimens are somewhat duller.

Below: *Culcita schmideliana*
The distinctive tubercles are clearly visible. Calcified, beige-brown areas of skin help to give the animal rigidity.

Fromia elegans
Red starfish

This bright red species is another common import in shipments from Indonesia and the surrounding region. It is very easy to keep, provided you buy undamaged specimens. In view of the sensitivity of the water vascular system, it is important to avoid rapid changes in salinity when acclimating starfishes to the aquarium.

   F. elegans is one of the smallest starfish species, reaching only 8cm(3.2in) in diameter. Juveniles have black tips to the arms, but these disappear as the animal matures. *F. elegans* requires a similar diet to *F. monilis* but is itself often eaten by more aggressive tankmates, such as hermit crabs and large starfishes.

Choriaster granulatus
Giant Kenya starfish

This species is probably the largest starfish available to the hobbyist on a regular basis. It can reach up to 30cm(12in) in diameter and is heavily built, with five thick fleshy arms. Unfortunately, this very attractive red and buff species is generally only imported when it has attained a size too great to appeal to the average hobbyist. This, and the proportionately high air freight costs, ensure that *Choriaster granulatus* remains a species for the specialist.

   The animal gets its scientific name from the small gill processes that protrude through the skin to give the arms a granular appearance. Most specimens are imported from Kenya, but it also occurs throughout the Indo-Pacific region, where it feeds on coral polyps and other immobile invertebrates.

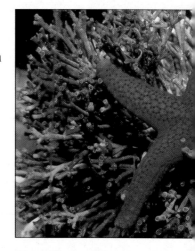

Above: *Fromia elegans*
A red starfish on an encrusting coral, *Xenia* sp., but neither this nor other *Fromia* species pose a threat to sessile invertebrates.

Right: *Fromia monilis*
A welcome splash of colour among the greens and browns of a well-stocked 'living reef' aquarium. Inexpensive, easy to maintain and justifiably popular starfishes.

Below: *Choriaster granulatus*
A giant starfish, probably eating the coral *Pachyseris rugosa* on which it is shown resting here.

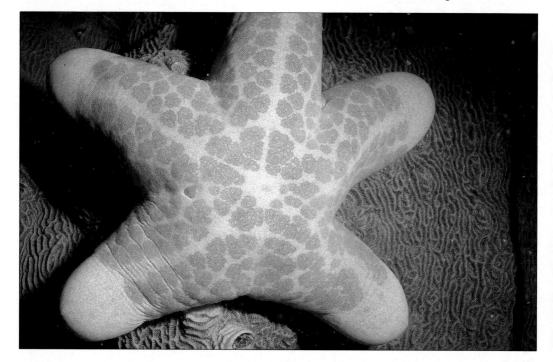

Fromia monilis
Orange starfish

This vivid orange and red species is one of the most popular and commonly imported starfish. Regular supplies from Sri Lanka and Indonesia ensure that they are within the financial reach of all marine hobbyists. They rarely grow more than 6cm(2.4in) in diameter, but can reach up to 10cm(4in).

Bear in mind that a starfish's water vascular system is easily damaged (see page 98), so examine all starfishes before you buy them to ensure that they are in good condition, particularly the tips of the arms, and that the body is not limp and flaccid. A starfish with any of these defects is unlikely to survive.

This species will not harm other invertebrates and feeds happily on small pieces of shrimp or shellfish. Highly recommended for the marine invertebrate tank.

Protoreaster lincki
Red-knobbed starfish

The red-knobbed starfish is a widely distributed Indo-Pacific species that can reach up to 30cm(12in) in diameter. It is mostly offered for sale at about half this size, and its long, vivid red dorsal spikes ensure that it makes an impact in the aquarium. The most attractive specimens have a red, netlike, pattern on an off-white background, and are quickly snapped up by hobbyists. Unfortunately, most collectors are not aware of a general rule that applies to starfishes, namely that knobbly backed starfish are omnivorous, if slow, predators, while most of the tropical, smooth-armed species are less harmful scavengers. This ignorance often results in red-knobbed starfish being introduced to an aquarium well stocked with corals, molluscs and other sessile animals, all of which seem to provide grist to this insatiable animal's mill.

In the right circumstances, red-knobbed starfish are rewarding and long-lived aquarium animals. Feed them by placing them directly onto 1.25cm(0.5in) pieces of fish, squid or shellfish.

Above: *Protoreaster lincki*
The tube feet can prise open most bivalve molluscs, while the stomach can be everted through the mouth to engulf sessile invertebrates.

Pentaceraster mammillatus
Common knobbed star

This species is extremely variable in colour. The background is mostly brown or green, and the knobs – which are substantially smaller than those of *P. lincki* – may be white, yellow, orange, brown or black. Most commonly seen at around 8cm(3.2in), they can reach twice this size.

As its name suggests, this is a common species, regularly included in imports from Singapore, the source of many of the cheapest invertebrates available to many hobbyists. Like its relative *P. lincki*, it is a greedy feeder, capable of everting the stomach in order to digest food. A suitable diet can include shredded prawn and brineshrimp; do not allow uneaten food to pollute the tank.

Right: *Pentaceraster mammillatus*
This is a common colour form of this most variable species. Bear in mind that starfishes with heavily knobbed arms are all likely to eat less mobile animals.

Left: *Protoreaster lincki*
The intense, cherry red patterning of this species has ensured its popularity within the hobby, but select tankmates with care.

Below: *Pentaceraster* sp.
The strong, brightly coloured, calcified knobs on the arms and body offer some protection from triggerfish and other predators.

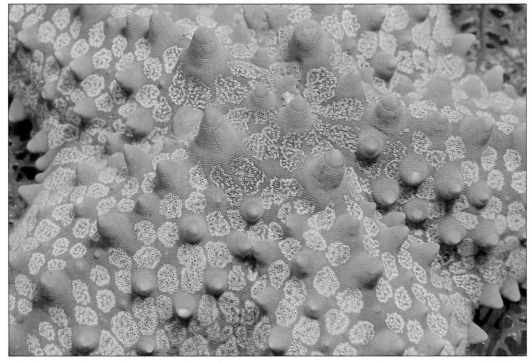

Echinometra mathaei
Common urchin

As its common name suggests, this Indo-Pacific species is not only widespread in the wild, but frequently available to hobbyists. Although its spines are shorter and considerably blunter than those of *Diadema savignyi* (page 181), take care when handling it.

E. *mathaei* is extremely easy to maintain in captivity; it is not too fussy about water conditions and will accept a wide variety of food. Unfortunately, it seems to spend a great deal of time hidden behind, or under, rocks and is most active at night. Bear in mind that the tube feet provide the urchin with a very strong grip on the substrate, allowing it to go almost anywhere it pleases, and in doing so it can easily tumble precariously positioned rockwork.

The shell of the common urchin can reach a diameter of about 10cm(4in), but most animals are about half this size and these make the better aquarium specimens.

Eucidaris tribuloides
Mine urchin

This small species only reaches about 5cm(2in) in diameter and, although not particularly attractive, it is a hardy species. It spends most of the day hidden in crevices in rocks, only emerging to feed at night and is included here to avoid confusion with the *Heterocentrotus* pencil urchins (see page 180).

Above: *Eucidaris tribuloides*
The mine urchin is a common export from Florida. It is shown here in an aquarium setting, but its nocturnal habits have restricted its popularity within the hobby.

Left: *Echinometra mathaei*
This common species uses its spines and tube feet to amble over the coral rubble of an Indonesian lagoon as its seeks out algae and detritus. A valuable scavenger in the aquarium and easy to feed.

Acanthaster planci
Crown-of-thorns starfish

The common name of this notorious species describes the animal admirably, both its looks and its degree of unpleasantness.

Native to the Indo-Pacific area, it can reach plague proportions and, in recent years, has caused considerable damage to many reefs, including the Great Barrier Reef, through its seemingly insatiable appetite for living coral polyps. Within a few hours of settling on a healthy coral head it will move on, leaving behind the dead, white skeleton.

Early attempts to control these plagues revealed a regrettable lack of understanding of starfish biology. Scuba divers were sent out with knives and machetes to hack the starfish apart. In many cases, this merely produced starfish fragments that regenerated as new animals and exacerbated the problem. It is still not known whether commercial fishing for the large univalve molluscs that prey on crown-of-thorns was responsible for the plagues, or whether they are part of a naturally recurring cycle, but the threat to Australia and Indonesia's reefs seems to have abated in recent years.

From this it is easy to draw the conclusion that *A. planci* is hardly suitable for the average aquarium and, as its spikes are capable of inflicting severe wounds and poisoning, it is clearly an animal to avoid.

Below: *Acanthaster planci* A voracious and dangerous species.

Heterocentrotus mammillatus
Pencil urchin

The pencil urchins of the family *Heterocentrotus* are generally considered the most attractive and desirable sea urchins but, sadly, they are neither as common in the wild nor as regularly imported as other species. This ensures that they command a price roughly twice that of most others. In *Heterocentrotus mammillatus* – and the closely related *H. trigonarius* – the spines are reduced in number, but those that remain are greatly thickened and resemble a 5-7.5cm(2-3in) pencil stub. Indeed, you can use these spines instead of chalk to write on a slate or stone tablet as did the Ancient Egyptians. Unfortunately, pencil urchins are often commercially fished to provide component parts for the wind chimes offered for sale in seaside gift shops.

Pencil urchins make admirable aquarium inhabitants, but be sure to keep the pH level of the water at the high end of the range. They eat algae, lettuce and small particles of meaty foods.

Above:
Heterocentrotus mammillatus
The distinctive heavy, thick spines
would seem to be something of an
encumbrance to the pencil urchin
as it roams across a Hawaiian reef.
Expensive, but easily maintained.

Left: *H. mammillatus*
The ventral spines, though smaller
than the dorsal, are still much
enlarged compared with those of
most other species. The banding
adds to the appeal of this specimen.

Right: *Diadema savignyi*
The long, needle-sharp spines
typical of this family of urchins are
clearly visible on this Indo-Pacific
species. They can cause painful
injuries; handle with care.

Diadema savignyi
Long-spined sea urchin

D. savignyi is just one member of a large family found throughout
tropical and subtropical seas. It may be uncomfortably familiar to
holidaymakers who have received painful wounds from treading on
these animals. The spines are long, extremely sharp and, in some
species, venomous. Despite this, they make good aquarium
inhabitants, grazing over algae-covered rocks and surviving for a
number of years. Their main drawback is that their sharp spines can
puncture and damage corals and sea anemones.

   D. savignyi is rather unusual among its family in that the dark
spines become lighter and banded at night. Most juvenile *Diadema*
species have banded spines as juveniles, but black or dark brown
spines when mature.

Toxopneustes pileolus
Poison urchin

At most, the average aquarist can expect a minor puncture wound
from his sea urchins, but a few species are positively dangerous
and one of these is *T. pileolus*. Here, the spines are greatly reduced
and the pedicellariae greatly enlarged and increased in number. In
most sea urchins, these organs are usually very small and
inoffensive, but in the venomous species they constitute efficient
three-jawed units armed with a poison that can produce a severe
reaction. They are used to capture and kill small prey animals, but
can be very painful to man.

   Fortunately, these species appear on the market very rarely but,
as in *Asthenosoma varium*, they are often an attractive reddish
orange colour, so a warning against them is merited.

Above: *Toxopneustes pileolus*
Its powerful venom is an effective
defence against predators.

Cucumaria miniata
Feather cucumber

Feather cucumbers may be among the smallest members of their family, but they are some of the most attractive and an ideal species for the newcomer to the hobby. There are several similar species, but the most common one, from Singapore, has a bright pink body and five longitudinal rows of bright yellow tube feet. In this species, the mouth tentacles are developed into feathery growths that the animal uses to trap planktonic organisms and other food items drifting in the water. They are very easy to feed with one of the commercially prepared liquid invertebrate foods, but are quite capable of trapping adult brineshrimp.

This and the following species are among the few that may successfully reproduce in a domestic aquarium. Where two or more of the 6cm(2.4in)-long adults are present you may find small clusters of miniatures adhering to the sides of the tank.

Synapta maculata
Worm cucumber

Exporters of invertebrates often include somewhat unusual animals in their shipments to make up the consignment when they run short of anemones, shrimps etc. Among those that appear from time to time are the worm cucumbers. These off-white, beige and brown creatures look rather like giant worms with feathery mouth parts. Like *Stichopus* sp. they ingest quantities of coral sand and extract what goodness they can, but they are much more active and flexible. They can move around the tank comparatively quickly and may be as much as 60cm(24in) long.

While interesting, they are hardly attractive and can be most unpleasant creatures, since their whole body is extremely sticky, clinging to fingers, rocks and anything else they contact.

Below: *Synapta maculata* A species best avoided.

Above: *Cucumaria miniata*
Yellow specimens of this small sea cucumber are less often seen than the pink form, which is regularly imported. Both are easy to keep.

Right: *Pseudocolochirus axiologus*
The sea apple has been a mainstay of the hobby for many years. The red tentacles are efficient traps for floating particles of food.

Below: *Pseudocolochirus* sp.
This vividly coloured Australian sea apple is in demand, despite its high price. Here, the feeding tentacles are semi-retracted.

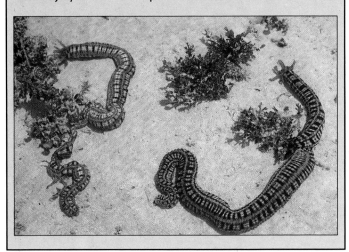

Pseudocolochirus axiologus
Sea apple

This very attractive species of sea cucumber has been given the somewhat confusing common name of 'sea apple' to differentiate it from all other cucumbers. It is by far the most strikingly coloured and popular species in the family.

The Indonesian form of *P. axiologus* typically has an ovate, greyish pink body up to 10cm(4in) long, with rows of tube feet defined in pink, orange or yellow. The head of the animal is crowned with a ring of feathery tentacles, which it uses for filter feeding. As the animal feeds, it pushes each tentacle in turn lugubriously into its mouth in the centre of the ring, rather like a schoolboy sucking toffee from his fingers. These tentacles vary in colour from pale yellow through to crimson. Do not keep sea apples with any fish species that might peck at the feathery tentacles.

P. axiologus is an easy species to maintain in the aquarium, provided it is given plenty of fine food. All too often, however, they slowly starve, becoming progressively smaller until, at about 3cm(1.2in) long, they give up the fight.

There is an even more striking giant form of *P. axiologus* from the Great Barrier Reef of Australia. Its body may be 15cm(6in) or more long, and is usually a rich purple colour with tube feet outlined in scarlet. The tentacles are purple and pure white. This desirable form is not cheap and beyond the pocket of many aquarists.

Stichopus chloronotus
Black cucumber

Sea cucumbers are a large, varied and largely unprepossessing family within the echinoderms. Looking like a dark, shrivelled and discarded cucumber, *S. chloronotus* is a fairly typical example of the group. It is found on coral rubble throughout the Indo-Pacific region where, like many of its relations, it swallows mouthfuls of gravel and detritus, digesting any organic material and ejecting the residue from its rear end.

This species is commonly available from Sri Lanka and, although very hardy, is not sufficiently attractive to appeal to most hobbyists. Nonetheless, it is a useful scavenger, particularly in an aquarium where there is a risk that organic material may begin to accumulate in a thin layer of unfiltered substrate.

Left: *Stichopus chloronotus*
Photographed here in shallow water off Queensland, this is one of the commonest Indian and Pacific ocean species. It is sometimes host to small fishes, *Carapus* sp. that seek refuge in the cloacal chamber.

Below left: *Didemnum molle*
One of many sea squirts found in large aggregations. They typically live in rather deep water, under ledges or within caves, where they are less susceptible to predation.

PHYLUM CHORDATA
SEA SQUIRTS

This group is mainly composed of animals with backbones, but members of two subgroups are considered as invertebrates, since they lack a true backbone. The sea squirts *Distomus* spp. (class Ascidiacea) are the only ones of interest to most hobbyists and even these usually arrive by accident. Small specimens are common introductions on pieces of living rock – most often white, beige or reddish species. They thrive with no additional care other than that given to the other inhabitants of the aquarium.

There are more than a thousand species of sea squirts. Some form mats of small specimens, others are large 50cm(20in) individuals. They have a leathery baglike body, with large inlet and outlet syphons. As water is drawn through these tubes, small particles of food are filtered out. Sea squirts are found in every colour and combination of colours, but their general inactivity means that they have never become as popular as might have been expected.

Above: *Pycnoclavella detorta*
Several of the smaller sea squirts, such as this semi-transparent species, live in tight colonies.

Right: *Cyclosalpa* sp.
In this large red species the inhalant and exhalant syphons (right and top) are clearly visible.

Below: *Rhopalaea crassa*
This attractive blue species is one of the most appealing sea squirts, but is only rarely available.

FISHES FOR THE INVERTEBRATE AQUARIUM

Very few hobbyists wish to keep an invertebrate aquarium totally devoid of fishes. A few fish within the tank add life, movement and colour, but you must exercise some caution when selecting suitable species. An ever-increasing range of fishes is available to the hobbyist and although some are suitable for inclusion with invertebrates, many others are not.

As a general guide to suitability, consider the diet and natural lifestyle of the fishes in question. Will they pose a threat to the invertebrates in your aquarium? In nature, we can say that food for fishes falls into one of three categories. In the first category are planktonic and particulate foods. In the second come the invertebrates, some of which we want to keep in the aquarium, but upon which many fishes feed. Finally, there are the small fishes that constitute food for many of the larger fish species.

Clearly, any fishes that rely on the first food source are likely to be safe with invertebrates, while the large predators, such as groupers, lionfishes and eels, will ignore corals, anemones, and sponges, etc., but may eat shrimps, crabs and some molluscs. Between these two extremes is a large group of omnivorous species with very catholic tastes. Some of these should never be included with invertebrates. These include triggerfishes, pufferfishes, large

wrasse species and all the butterflyfishes, with the possible exception of the longnosed species *Forcipiger longirostris* and *Chelmon rostratus*. These two species take small worms and crustaceans from among coral heads, rather than eat the corals themselves, which is common among other butterflyfishes (*Chaetodon* spp.).

If you want to keep the full spectrum of invertebrates, you will have to limit yourself to the smaller fish species that feed on plankton, etc. These include damselfishes, clownfishes, dwarf groupers, gobies and the various blennies. However, as long as you do not keep small crustaceans, then lionfishes, large groupers, ribbon eels and some of the smaller moray eel species make dramatic tank inhabitants.

Unfortunately, many of the most attractive marine fish are omnivores and two main considerations apply before you introduce these species. Firstly, many of them will peck at – and eat – the feathery tentacles of filter-feeding invertebrates. Featherduster worms, sea cucumbers and some soft corals are particularly prone to this form of predation. Secondly, some species may cause considerable accidental damage. A fish can only investigate a possible food source with its mouth. After taking one bite of, say, a coral, it may decide that it has had enough, but may cause sufficient damage to result in the invertebrate's eventual demise.

We now consider a selection of suitable fishes for the invertebrate aquarium, with appropriate mention of some to be avoided.

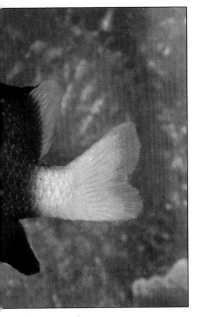

Above: *Dascyllus carneus*
The dusky damselfish is one of the smaller *Dascyllus* species. It is more peaceful than *D.trimaculatus*, the common domino damselfish.

Left: *Amphiprion ocellaris*
The common clownfish is an ideal choice for the invertebrate aquarium with anemones. A single fish often defends its anemone from intruders.

Below: *Pomacentrus coeruleus*
Electric blue damsels make an inexpensive but colourful impact in the 'living reef' aquarium. Easily maintained, if a little boisterous.

Clownfishes
Amphiprion and *Premnas* spp.

These attractive red, pink or orange-and-white species are justifiably popular. They are famous for their symbiotic relationship with sea anemones, whereby they acquire an immunity to the anemone's sting and are able to live among the tentacles, thus avoiding predators. There are numerous species; most are readily available and all are easy to maintain. Small clownfishes start life as males, with the dominant specimen becoming female. To acquire a pair, simply buy two small specimens and let them grow. Clownfishes commonly spawn in the aquarium, but rarely manage to raise their young.

Damselfishes
Abudefduf, Chromis, Dascyllus, Pomacentrus spp. et al

Damselfishes are among the cheapest marine fish and, being hardy, will tolerate less than perfect water conditions. Many are attractive as youngsters, but turn a dull grey or brown colour as they mature. The majority also have a reputation for becoming aggressive towards other fish as they grow older. As a general rule, the blue species tend to be more peaceful than the black/white/yellow combinations. Damsels are generally safe with all but the smallest crustaceans. They will accept any form of food and may spawn in the aquarium, usually on coral branches or shells.

Dwarf groupers
Pseudochromis and *Gramma* spp.

Few of these small fishes grow more than 10cm(4in) long. Of all the fishes suitable for the invertebrate tank, these are some of the jewels and, in their dazzling combinations of red, purple, yellow, orange, blue, gold and black, merit a place in every tank. *Gramma loreto* (purple and yellow) is one of the best. It is easily obtainable, peaceful and occasionally spawns, lining a small crevice with algae to make a nest. Unless you have a very large tank, keep only one specimen of a species.

Wreckfishes; reef-fishes
Anthias and *Mirolabrichthys* spp.

Numerous species of *Anthias* (also members of the grouper family) live in large shoals on the Indo-Pacific reefs, where they feed on planktonic organisms. There is considerable sexual dimorphism; the males are frequently reddish or green, while the females of many species are orange-yellow. These very placid fishes are best kept in small groups of one male with several females. They feed well on brineshrimp and *Mysis* shrimp, but demand frequent feedings and good water quality. They can be trusted with any invertebrate likely to be kept in the aquarium.

Right: *Pseudochromis dutoiti*
Neon-backed basslets are a large *Pseudochromis* species. *P. dutoiti* may eat small shrimps, but is harmless to sessile invertebrates.

Below right: *Anthias squamipinnis*
This wreckfish is easy to obtain and ideal for the invertebrate tank. Keep a small shoal together.

Below: *Gramma loreto*
In a mixed invertebrate and fish tank, the royal gramma is safe with all but the smallest crustaceans.

Gobies and blennies

This loose grouping includes many distinct families, but for practical purposes they can be considered under one heading. The vast majority are small, bottom-dwelling fishes with interesting, rather than strikingly coloured, patterns. However, there are numerous exceptions and some of these are the most popular fishes for invertebrate tanks. Here, we consider mandarinfishes, firefishes and eyelash gobies.

Mandarinfishes
Synchiropus spp.

These vivid, frog-faced little fishes feed from the substrate and not in midwater, so do not keep them with more competitive feeders. They are particularly useful in eating the small crustacean copepods that can irritate fishes and decimate decorative algal growths. Males are distinguished by the extended first ray of the dorsal fin.

Below: *Synchiropus splendidus*
Mandarinfishes are useful and very attractive scavengers that will eat troublesome parasitic copepods.

Right: *Escenius* spp.
Most species of eyelash blennies are peaceable and entertaining additions to the aquarium.

Firefishes; fire gobies
Nemateleotris spp.

These species, with their delicate colours – they can be pink, purple, yellow or white – are strongly recommended. In the early 1970s they were considered rarities, but today they are readily available and a small shoal of the aptly named firefish is delightful. Although somewhat shy initially, they soon settle down to become hardy, easily kept and totally peaceful tank inhabitants. In recent years, several new species have appeared on the market, all of which suit the 'living-reef' set-up.

Eyelash blennies
Escenius spp. and others

These generally small (up to 7cm/2.75in)-long eel-like fishes are extremely hardy and ideal for the beginner. They will accept any commercially available food that will fit their rubber-lipped mouths and are useful in eating the encrusting green algae that forms on rocks. As their common name suggests, they sport a series of growths above the eyes, reminiscent of eyelashes. The orange-tailed *Escenius bicolor* is one of the most commonly available species. During spawning, males turn red with white bars.

Many other species of gobies make ideal aquarium inhabitants, including the blue, black-and-white-striped neon goby and the scarlet-and-blue catalina goby.

Below: *Nemateleotris decora*
The purple firefish is attractive, hardy and totally inoffensive to invertebrates and other fishes.

Cardinalfishes
Apogon spp.

These 7cm(2.75in)-long fishes are more popular and readily available in the United States than in Europe, but are ideal for the invertebrate aquarium. They do best in small groups, single specimens often proving shy and retiring. The most attractive species are vivid red, but the common spotted species, *Sphaeramia* (*Apogon*) *nematopterus*, has a certain charm. Cardinalfishes are mouthbrooders, in common with many of the freshwater tropical cichlid species.

Seahorses and pipefishes
Hippocampus and *Dunkerocampus* spp.

Seahorses and their relations fare much better in an invertebrate aquarium than in a fish-only system, but you must give some careful thought to selecting suitable tankmates. They are particularly weak swimmers, and it is not unusual for them to blunder into anemones, only to be stung and digested. Some of the larger, more heavily clawed crustaceans, such as *Stenopus* shrimps and hermit crabs, are capable of catching and killing these fishes.

Above: *Sphaeramia nematopterus*
Keep these cardinalfishes in groups of three or more, otherwise they may prove shy. The larger, related species are more aggressive and liable to eat crustaceans.

Left: *Hippocampus kuda*
Seahorses are delicate aquarium subjects. Large sea anemones and crustaceans sometimes prey on them and, being totally placid, they cannot tolerate competition from more lively fish species. Slow death by starvation often results.

Below: *Syngnathus dactyliophorus*
The banded pipefish is closely related to the seahorse. The male carries eggs under the belly. Pipefishes need careful feeding with small livefoods and should be left to experienced aquarists.

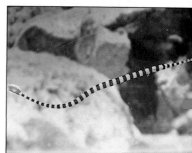

Furthermore, they are particularly slow feeders and are all too often housed with much more competitive fish species. In these circumstances, a seahorse may survive for a few weeks, but eventually it will succumb to starvation, one of the commonest causes of death among seahorses.

The larger, readily available Indo-Pacific seahorses are generally easier to keep than the dwarf Caribbean species. Initially, they may demand livefood in the form of adult brineshrimp, but they usually learn to take frozen brineshrimp and *Mysis* shrimp, if not subjected to excessive competition from other fish species.

Hawkfishes
Cirrhitichthys and *Oxycirrhites* spp.

Hawkfishes are ideally suited to the invertebrate aquarium, as they are harmless to all but very small crustaceans, are very hardy and easily maintained, and many species are inexpensive. Hawkfishes have very poorly developed swimbladders and, as one common name suggests, appear to hop from rock to rock. They become very tame, quickly learning to recognize their owner, and easily merit their growing popularity.

The two most desirable species are the longnosed hawkfish, *Oxycirrhites typus*, and the flame hawkfish, *Neocirrhites armatus*. The former has a red and white chequered pattern and a greatly extended jaw structure. The latter has the most intense red colouring, highlighted by a velvet black stripe down the back. Few of the hawkfishes grow longer than 10cm(4in), but those few species that do will occasionally eat small fishes.

Below: *Paracirrhites forsteri*
The spectacled hawkfish is harmless to sessile invertebrates, but is an eager predator, taking small shrimps and crabs. Smaller hawkfishes pose less of a threat.

Dwarf angelfishes
Centropyge and *Geniacanthus* spp.

Among the many dwarf angelfishes are some of the most attractive species available within the hobby, and several are suitable for the invertebrate aquarium. However, their naturally omnivorous diet must be taken into consideration. They will almost invariably peck at feathery-tentacled invertebrates and will eat quite large amounts of what the aquarist may consider decorative forms of algae.

As a general rule, the smaller the species the less damage they will cause. Thus, *Centropyge acanthops*, *Centropyge argi* and *C. resplendens*, which only reach about 6cm(2.4in) long, can be trusted in most aquariums. The larger species are generally safe with anemones, clams, starfishes and crustaceans, but may on occasion damage corals. The most desirable of these larger species is the flame angel, *C. loriculus*, which grows to about 10cm(4in) and is a vivid orange-red with black vertical bands. Although these are some of the most expensive dwarf angels, their good manners in invertebrate aquariums justifies their popularity.

Most *Geniacanthus* species are fairly well behaved, but as many reach 15cm(6in) or more in length and are active swimmers, do not include them in a small tank, i.e. less than 120cm(48in) long.

(The large angelfishes of the *Pomacanthus*, *Holacanthus* and *Euxiphipops* families are equipped with strong, beaklike jaws that can rapidly inflict a lot of damage. Although individuals vary greatly in temperament, do not, as a general rule, house them with invertebrates except, possibly, large anemones and crustaceans.)

Above: *Geniacanthus melanospilus*
The zebra angelfish is one of the largest dwarf angels. It may peck at 'feathery' invertebrates, but rarely causes significant damage.

Right: *Pterois volitans*
Lionfishes can reach 25cm(10in) or more and make a dramatic centrepiece in a large aquarium. Safe with many invertebrates.

Below: *Centropyge loriculus*
The vivid coloration and peaceful nature of the hardy flame angelfish combine to ensure its continued popularity with marine hobbyists.

Above: *Dendrochirus zebra*
These interesting, if inconspicuous additions to the invertebrate tank eat small crustaceans and fish but will not harm sessile animals.

Lionfishes
Pterois and *Dendrochirus* spp.

The suggestion that lionfishes are suitable for invertebrate aquariums often produces raised eyebrows, but since their natural diet consists of small fishes and shrimps, they pose no threat to corals, anemones, starfishes, urchins and the wide range of sessile animals available today.

All members of this family are venomous, being armed with a row of needle-sharp spines down the back. Some species have additional spines around the gill covers. These spines are capable of injecting a cobralike venom into the fingers of an unwary aquarist, causing acute pain and, occasionally, a much more severe reaction. First-aid is very simple and effective; immerse the wound in hot water for not less than five minutes. The heat breaks down the protein in the venom and reduces its effect, but further medical attention may be necessary.

Despite these warnings, remember that lionfishes only sting in self defence. They much prefer to avoid confrontation, so any stings can usually be blamed upon careless fishkeepers.

Eels
Muraenia and *Rhinomuraenia* spp.

Like the preceding family, the various moray eels and the ribbon eel can be trusted with sessile invertebrates. However, they are eager predators of small fishes and crustaceans, and large specimens are well equipped with razor-sharp teeth. In suitable circumstances, they make dramatic additions to an invertebrate aquarium.

Wrasses
Coris, Cirrilabrus, Pseudocheilinus, Macropharyngodon spp.

The wrasses include some of the smallest and the largest marine fishes, with adult specimens varying from 5cm(2in) to over 2m(6.6ft) in length. The larger species – *Bodianus, Thalassoma* and the bigger *Coris* – can be very destructive and have no place among invertebrates. However, there are many small species that are easy to maintain and make well-behaved inhabitants of 'living-reef' aquariums. In recent years, the *Cirrilabrus* wrasses have come to the fore, with a seemingly endless supply of new and brightly coloured species becoming available. These grow to about 10cm(4in) and all can be heartily recommended.

Above:
Rhinomuraenia amboinensis
The ribbon eel can easily escape.
Provide a secure lid on the tank.

Above right: *Lactoria cornuta*
The long-horned cowfish can occasionally cause damage by chewing on featherduster worms.

Left: *Cirrilabrus* sp.
The Maldive parrot wrasse is one of many *Cirrilabrus* species. All are long-lived, hardy and peaceful.

Below: *Anampses rubrocaudatus*
The red-tailed wrasse is one of the most attractive Hawaiian fishes. It is generally trustworthy with all but the smallest crustaceans.

Boxfishes
Ostracion, Tetrasomus and *Lactoria* spp.

The various boxfishes and their close relations are frequently kept among invertebrates but, in common with surgeons and tangs, *Acanthurus* and *Zebrasoma* spp., they tend to suffer from white spot. Larger specimens often prove unacceptably destructive, killing corals and starfishes. Even the smallest specimens generally attack and kill featherduster worms and sea cucumbers. Under extreme stress, boxfishes are capable of producing a poison, which may result in the death of all the fish in the tank. Fortunately, most invertebrates are not seriously affected by this toxin.

MARINE ALGAE

To freshwater hobbyists coming to marines for the first time, the mention of 'algae' is likely to conjure up a picture of greenish brown slime growing over plants and masses of hairlike tendrils choking the aquarium. By contrast, marine-keepers consider the different forms of algae as friends and allies.

The many species of algae that live within the tissues of corals and anemones – the zooxanthellae – provide food and help with the elimination of the animals' waste products. Small encrusting, and generally insignificant, algae coat the rockwork, producing a more natural-looking scene and providing a continually available food source for many browsing invertebrates and fishes. The larger species include many decorative forms, and these are the marine equivalent of aquarium plants in a freshwater tank.

The encrusting species and those that form furry 'lawns' on rocks generally arrive as accidental introductions with living rock, or as small particles included in the water when you buy invertebrate specimens. The larger, decorative species of algae are usually bought separately, but some are so prolific that you may be able to obtain 'cuttings' from neighbouring aquarists. They may be green, brown or red in colour.

Like most terrestrial plants, all algae species use chlorophyll to synthesize food, so require moderate to strong lighting. Commercial algae fertilizers are available, but these are only rarely necessary. Indeed, if other conditions are less than ideal, they can do more harm than good, by promoting undesirable species and increasing nitrate levels within the tank. Generally speaking, sufficient trace elements are dissolved within the water, and enough phosphates and nitrates are available from the animals' wastes to ensure good algal growth. In fact, by absorbing nitrates and phosphates as plant fertilizers, all forms of algae play an important role in maintaining good water quality in the aquarium.

In the early days of the newly established invertebrate aquarium, you may encounter a problem with excessive growth of unwanted or undesirable forms of algae. This is most common where high intensity lighting is provided and mats of green, hairlike algae form as a result. Generally speaking, the sequence of events is as follows. The aquarist provides sufficient light for zooxanthellae algae to function but, when the tank is new, large expanses of rock are exposed to this light. As there are few corals and anemones to utilize the light, algae is able to prosper. However, as soon as more corals and similar invertebrates are added to the tank, they begin to shade the algae and use the light themselves, eventually causing the algae to decline. Sea urchins and cowries will graze on green algae, but these animals are somewhat indiscriminating about the routes they follow and may damage sessile animals in the process.

Although most marine algae can be safely encouraged to grow, there is a purple-brown alga that rapidly forms a spreading film over everything in the tank. Be sure to siphon any spots of this type of alga out of the tank as soon as you notice them. This problem is normally associated with overstocking, incorrect lighting, overfeeding, insufficient water movement within the tank,

Above: The varied colours and forms of marine algae add an extra and welcome dimension to a tank of hard and soft corals. Most *Caulerpa* species are easy to grow.

insufficient frequency (or quantity) of water changes, and excessive levels of nitrates. A series of partial water changes and an improvement in general conditions will usually solve the problem. On no account should you use the algae-killing preparations designed for ponds and freshwater aquariums in a marine system.

Caulerpa spp.

The most commonly cultivated algae are the many *Caulerpa* species. Various forms are found throughout the Caribbean, Mediterranean and Indo-Pacific regions and all are prolific. *Caulerpa* species may vary in colour from an intense lime green to a bluish brown, and they may grow 50cm(20in) tall or form low mats. Despite this variation, the basic structure is very similar. Growth develops from a main runner, with leaf stalks being produced from the top of the runner and a rootlike growth from the bottom. This 'holdfast' serves to anchor the plant body in position. It does not absorb water and food in the same manner as a terrestrial plant's roots; in marine algae, nutritional substances are absorbed through the leaves.

The fronds of *Caulerpa* species are very thin-walled and filled with fluid. Because of this, it is important to acclimate these plants very slowly to a new environment, especially in terms of specific gravity. If the transition is too sharp, changes in osmotic pressure can rupture the cell walls, causing an attractive green plant to change rapidly into a decomposing translucent slime.

Since some 'body fluid' will leak when pieces are removed for transplanting to other aquariums, it is safer to buy larger, rather than smaller, segments to reduce the proportion lost. This is particularly important if the tank houses fishes or invertebrates that may peck at the plant. A large sample has a much better chance than a small piece of surviving – and outgrowing – this minor pruning process.

In some circumstances, *Caulerpa* species grow so rampantly that they threaten to swamp, or at least severely shade, the various corals and polyps in the tank. Regular thinning of the growth is much more desirable than infrequent but heavy pruning, which may, on occasion, cause the collapse of the whole growth through excess fluid loss from the cut surfaces.

Most *Caulerpa* species are easy to grow, although the fleshier species require greater care than the more leathery types. The species can be distinguished by the form of the 'leaves'. The easiest to identify is probably *C. prolifera*. It has a thin, wiry runner, or stolon, typically up to 30cm(12in) tall and straplike leaves up to 3cm(1.2in) wide. Occasionally, the leaves develop as chains of heart-shaped sections, usually as a result of persistent damage to the growing tips of the leaves.

Caulerpa mexicana and *C. seratuloides* have very attractive featherlike leaves and both grow very rapidly. *C. racemosa* has short bunches of spherical or ovate, berrylike growths on the vertical stalks and has earned the species the common name grape caulerpa. Although very attractive, it is rather slow growing and somewhat more demanding than most species.

Below: Some growth of higher algae is desirable, but regular pruning is vital if sessile animals, such as these *Zoanthus* polyps, are not to become overgrown.

Codiacea spp.

Several attractive members of this algae group are imported, primarily from the Caribbean. These species are known as calcified algae because their 'leaves' are reinforced with calcium, absorbed from the surrounding sea water. This makes them much more rigid than *Caulerpa* species and less prone to predation. Given good lighting, a high pH level and regular use of a pH buffer solution, most species are easy to maintain, though often slow to reproduce.

Pencillus capitatus
Shaving brush

This is probably the most commonly imported of the Caribbean *Codiacea* and grows on a wide variety of substrates, its fleshy stem often buried 7cm(2.75in) deep in the sand or mud. Its common name is very apt, as it perfectly resembles a green shaving brush. Although easily damaged in transit, intact specimens usually flourish, with small new growths appearing like rose suckers. *Rhipocephalus phoenix* is similar in appearance, but has a crown made up of concentric rings of thin, flattened plates.

Halimeda spp.

Typical specimens of these algae consist of a stalk, anchored into a soft substrate, from which grow numerous roughly circular or heart-shaped flat plates. New plates grow from the tips of the old ones. *H discoidea* is one of the commonest species, its numerous 1.5cm(0.6in)-diameter plates giving the impression of a prickly pear cactus – hence the common name cactus algae. *H. goreaui* and *H. opuntia* form very dense mats of tiny plates and look particularly attractive in a small aquarium. *H. copiosa* produces long chains of small plates and is very delicate and elegant.

Avarainvillea nigricans

This species is related to the two previous species, but here the stalk is topped by a single, large, flattened blade, roughly circular and up to 10cm(4in) in diameter. Although hardy, the blades are often coated with brown encrusting algae, which detract from the plant's appeal. *Udotea* species are very similar in appearance and, from the aquarist's point of view, can be considered one and the same.

Acetabularia spp.
Mermaid's cup

This small and very delicate plant is one of the most attractive of the marine algae, but it is only rarely available to the hobbyist. The thin stalks, only 5cm(2in) long, are topped by pale blue-green caps like inverted toadstool heads. Unfortunately, the plant is easily

damaged, both in transit and by other aquarium inhabitants, and is easily swamped by hairlike algae. It requires good light and less water movement than suits most invertebrates and algae.

Rhodophyceae
Red algae

Several decorative species of maroon-red algae are occasionally available. Typically, these are anchored to a base rock by a thick stem that rapidly branches to form a bushlike structure. Some species are quite stiff and erect, while others collapse if removed from water. The success rate with these types of algae is very variable. The best specimens are those that remain attached to a small rock and have few or no pale or faded tips to the branches.

Valonia ventricosa

This species produces a cluster of roughly spherical balls up to 5cm(2in) in diameter. Each ball is a single cell and it is this plant's claim to fame as the largest single-celled growth in the world that earns it a place here. *V. ventricosa* occurs as an accidental introduction into the aquarium and, given time, can make an attractive feature. The cells are easily punctured, however, and should be handled with great care.

Left: *Halimeda copiosa*
The slow but steady growth rate of *Halimeda copiosa* suits even the smallest aquarium. It is one of the most decorative cactus algae.

Below left: *Peyssonnelia squamaria*
This is a typical example of the form and colouring of several of the plate-forming calcareous algae. They often occur on 'living rock'.

Top left: *Codiacea* spp.
This attractive Red Sea algae shows the fleshy, fingerlike form typical of the family. They all need slow and careful acclimatization.

Below: *Valonia ventricosa*
The 'bunch of grapes' formation of this distinctive algae is clearly visible here. It appears in the tank as an accidental introduction.

INDEX

Page numbers in **bold** indicate major references including accompanying photographs. Page numbers in *italics* indicate captions to other illustrations. Less important text entries are shown in normal type.

ACKNOWLEDGEMENTS

The publishers wish to thank the
following individuals and organizations
for their help in the preparation of this
book:

Dr. M.G. Abeywickrama (Thorn
Lighting Ltd.); Terry Evans (Wet Pets);
Adrian Exell (Interpet Ltd.); Dick Mills;
David Saxby.

PICTURE CREDITS

FURTHER READING

Banister, K. and Campbell, A. *The Encyclopedia of Underwater Life* George Allen and Unwin, 1985.

Bemert, G. and Ormond, R. *Red Sea Coral Reefs* Kegan Paul International, 1981.

Debelius, Helmut *Armoured Knights of the Sea* Kernan Verlag, 1984.

Debelius, Helmut *Fishes for the Invertebrate Aquarium* Kernan Verlag, 1986.

George, David and Jennifer *Marine Life - An Illustrated Encyclopedia of Invertebrates of the Sea* Harrap, 1979.

Kaplan, E.H. *A Field Guide to Coral Reefs of the Caribbean and Florida* Peterson Field Guide 27. Houghton Mifflin, 1982.

Mills, Dick *A Fishkeeper's Guide to Marine Fishes* Salamander Books, 1985.

Mills, Dick *The Practical Encyclopedia of the Marine Aquarium* Salamander Books, 1987.

Straughan, R.L. *Keeping Live Corals and Invertebrates* A.S. Barnes & Co. Inc., 1975.